Professional Practice in Earth Sciences

Series Editor

James W. LaMoreaux, Tuscaloosa, AL, USA

Books in Springer's Professional Practice in Earth Sciences Series present state-of-the-art guidelines to be applied in multiple disciplines of the earth system sciences. The series portfolio contains practical training guidebooks and supporting material for academic courses, laboratory manuals, work procedures and protocols for environmental sciences and engineering. Items published in the series are directed at researchers, students, and anyone interested in the practical application of science. Books in the series cover the applied components of selected fields in the earth sciences and enable practitioners to better plan, optimize and interpret their results. The series is subdivided into the different fields of applied earth system sciences: Laboratory Manuals and work procedures, Environmental methods and protocols and training guidebooks.

More information about this series at http://www.springer.com/series/11926

Dilber Uzun Ozsahin · Hüseyin Gökçekuş ·
Berna Uzun · James LaMoreaux
Editors

Application of Multi-Criteria Decision Analysis in Environmental and Civil Engineering

 Springer

Editors
Dilber Uzun Ozsahin
DESAM Institute
Near East University
Nicosia, Turkish Republic of Northern
Cyprus, Turkey

Department of Biomedical Engineering
Near East University
Nicosia, Turkish Republic of Northern
Cyprus, Turkey

Medical Diagnostic Imaging Department
College of Health Science
University of Sharjah
Sharjah, United Arab Emirates

Berna Uzun
DESAM Institute
Near East University
Nicosia, Turkish Republic of Northern
Cyprus, Turkey

Department of Mathematics
Near East University
Nicosia, Turkish Republic of Northern
Cyprus, Turkey

Hüseyin Gökçekuş
Faculty of Civil and Environmental
Engineering
Near East University
Nicosia, Turkish Republic of Northern
Cyprus, Turkey

James LaMoreaux
P. E. LaMoreaux & Associates, Inc.
Tuscaloosa, AL, USA

ISSN 2364-0073   ISSN 2364-0081   (electronic)
Professional Practice in Earth Sciences
ISBN 978-3-030-64767-4   ISBN 978-3-030-64765-0   (eBook)
https://doi.org/10.1007/978-3-030-64765-0

This Springer imprint is published by the registered company Springer Nature Switzerland AG
The registered company address is: Gewerbestrasse 11, 6330 Cham, Switzerland

Contents

Chapter 1
Introduction

**Dilber Uzun Ozsahin, Berna Uzun, Aizhan Syidanova,
and Mubarak Taiwo Mustapha**

Abstract The application of Multi criteria decision making (MCDM) has sprung many facet of life. Its application in the field of civil engineering and environmental studies owe to fact that decision-makers in these fields are always in dilemma when confronted with challenges involving multiple criteria. In civil engineering, decision-making is critical to the success of any project. Any wrong decision can be detrimental not only to people's life but to the cost and quality of time spent on a project. Civil engineers are usually confronted with alternatives whenever a project is to be executed. This alternatives include the type, length and strength of a material to be used or its longevity. Similar alternatives are peculiar to environmental studies. Climate change has become the most debatable topic since the peak of industrial revolution. Greenhouse gasses (carbon-dioxide (CO_2), water vapor and methane) has been emitted in an uncontrollable manner resulting in the damage of the protective ozone layer. This has led to a far reaching consequences such as drought, heat waves,

D. Uzun Ozsahin · B. Uzun (✉)
DESAM Institute, Near East University, Nicosia, Turkish Republic of Northern Cyprus, Turkey
e-mail: berna.uzun@neu.edu.tr

D. Uzun Ozsahin
e-mail: dilber.uzunozsahin@neu.edu.tr

D. Uzun Ozsahin · M. T. Mustapha
Department of Biomedical Engineering, Near East University, Nicosia, Turkish Republic of Northern Cyprus, Turkey
e-mail: mustaphataiwo54@gmail.com

D. Uzun Ozsahin
Medical Diagnostic Imaging Department, College of Health Science, University of Sharjah, Sharjah, United Arab Emirates

B. Uzun
Department of Mathematics, Near East University, Nicosia, Turkish Republic of Northern Cyprus, Turkey

A. Syidanova
Department of Architecture, Near East University, Nicosia, Turkish Republic of Northern Cyprus, Turkey
e-mail: aizhansyidanova@gmail.com

© The Author(s), under exclusive license to Springer Nature Switzerland AG 2021
D. Uzun Ozsahin et al. (eds.), *Application of Multi-Criteria Decision Analysis in Environmental and Civil Engineering*, Professional Practice in Earth Sciences,
https://doi.org/10.1007/978-3-030-64765-0_1

shrinking of the glacier ice, bush burning, deforestation etc. To provide a solution to these, several environmental friendly alternative needs to be consider.

Keywords Multi criteria decision making · Fuzzy logic · Environmental science

Multi-criteria decision-making analysis uses mathematical simulation tool to evaluate and compare conflicting alternatives based on multiple criteria. This result in ranking of alternatives from the most preferred to the least preferred. Asides criteria, weight and preference function are considered. In this textbook, the most popular and frequently used MCDM tools will be discussed and used. These are Analytic Hierarchy Process (AHP), Technique for Order of Preference by Similarity to Ideal Solution (TOPSIS), Elimination Et Choix Traduisant la Realite (ELECTRE), Preference Ranking Organization Method for Enrichment Evaluation (PROMETHEE) and Vise Criterion Optimization and Compromise Solution in Serbian (VIKOR). Fuzzy logic and fuzzy based MCDM will also be elaborated on.

Multi-Criteria Decision-Making (MCDM) has become a fairly well-known sector of the Operational Research/Management Science discipline; indeed, in recent years, numerous other methodologies and paradigms have been created in this area, with increasing use every day to a rather different type of problem scenarios.

MCDM methods have every chance of providing tools for difficult planning exercises, in terms of structuring values, weighing and choosing other possibilities, and distributing resources between competing for work styles. These methods still can take into account in contexts in which a certain number of interested stakeholders are involved, which in fact has the potential to lead to conflict situations. The use of MCDM techniques has flourished in every scientific field. Taking into account the distribution of these layouts and their variants, it is important to have an absolute awareness of their comparative value and ease of use in all kinds of contexts. As such, this special issue has a collection of methodological and applied cases that provide important merit in the field of making Multi-Criteria Decision-Making.

MCDM and MODM (Multi Objective Decision Making) it is two fundamental techniques of MCDM. The tasks of MCDM differ from the complication of MODM, which include producing the "leading" alternative by considering mutual concessions within a set of interacting design limitations. The MCDM method selects some areas of action if there are several, as a rule, incompatible attributes.

MCDM is an exceptionally famous branch of decision-making. This is a branch of a joint class of models for the study of operations that find solution of the troubles of a conclusion in the presence of a number of criteria for making a conclusion. The MCDM alignment urges that choice be made between the conclusions described by their attributes. It is expected, in fact, that the difficulties of MCDM have in advance a concrete, limited number of other conclusions. Conclusion of difficulty MCDM incorporates sorting and ranking.

In the MODM approach, contrary to the MCDM approach, the decision alternatives are not given. Instead, MODM provides a mathematical framework for designing a set of decision alternatives. Each alternative, once identified, is judged by how close it satisfies an objective or multiple objectives. In the MODM approach, the number of potential decision alternatives may be large. Solving a MODM problem involves selection.

The MODM approach differs from the MCDM approach in that no different solutions are provided. MODM demonstrates the mathematical basis for developing other conclusions. Any candidacy, once concrete, is evaluated by how close it meets the goal or a huge number of goals. In the MODM scenario, the number of probable other conclusions has the potential to be tremendous. Conclusion MODM difficulties imply choice.

It is widely recognized that the bulk of the conclusions adopted in the real world are accepted in an environment in which goals and limits due to their difficulties are not literally popular, and thus the problem does not have the ability to be literally defined or literally presented in exact form. Zadeh (1965) proposed using the concept of fuzzy sets as a modelling tool for difficult systems that have every chance of being controlled by people, but which are difficult to literally qualify to deal with high-quality, inaccurate information or even poorly structured conclusion problems (Bellman and Zadeh 1970).

Fuzzy logic is a section of arithmetic that allows programs on a computer to simulate the real world, the same world people live in. This is a simple method to reason with uncertain, diverse and inaccurate data or knowledge. In Boolean logic, any statement is considered true or false; that is, it contains the true meaning of 1 or 0. Numerous Booleans impose strict membership requests. Vague large numbers have more flexible membership requests that allow selective membership in the kit. Everything depends on the degree, and clear reasoning is considered as a limiting case of indicative thinking. Therefore, Boolean logic is considered a subset of fuzzy logic. People take part in the analysis of conclusions because the adoption of conclusions must take into account the subjectivity of a person, and not only apply impartial probabilistic measures. This prepares the adoption of fuzzy conclusions important. (Kahraman 2008).

The Technique for Order of Preference by Similarity to Ideal Solution (TOPSIS) is a technique that seeks to find the closest possible solution to the positive ideal solution (PIS) in a multi-criteria decision environment. It has many benefits. It's easy to use and organized. It has been used in supply chain management, logistics, construction, engineering and manufacturing systems, business and marketing successfully (Balioti et al. 2018).

From the Serbian language, *VIsekriterijumska optimizcija i KOmpromisno Resenje* (VIKOR) is a way of finding a compromise ranking created by Serafim Oprikovic. VIKOR is a method that determines the superior value in comparing two alternatives for a final set of other actions that must be ranked and selected between the criteria, and resolves a discrete multicriteria problem with disparate and conflicting aspects. VIKOR pays more attention to demanding and choosing one of the best from the set of variables and determines compromise difficulties with conflicting aspects that can help decision makers to show the final verdict. A compromise conclusion is the final conclusion among the alternatives, closer to impeccable (Lee and Yang 2017).

The VIKOR and TOPSIS methods are based on distance calculation, but the compromise conclusion in VIKOR is guided by mutual concessions, while in TOPSIS the best conclusion is guided by the minimum distance from PIS and the farthest distance from NIS (negative ideal solution). PIS is considered to be a type that consists of the best ratings between all considered criteria or attributes. On the other hand, NIS is considered a candidate that contain the worst ratings between all the criteria considered (Lee and Yang 2017).

The Analytical Hierarchy Process (AHP) was developed by Saati in 1980. AHP is an additive weighting method. It has been reviewed and used in many fields, and its implementation is maintained by several commercially available, user-friendly software packages. It is generally difficult for people who accept conclusions to literally qualify weights of total importance for a set of characteristics at the same time. As the number of characters' increases, the best results are obtained when the problem is transformed into one of a series of matched analogs. AHP formalizes a change in the difficulty of weighting characteristics into a more manageable problem of making a series of pairwise comparison between competing characters. AHP summarizes the results of matched analogs in the "matrix of paired comparisons". For any pair of attributes, the person accepting the conclusion reveals the outcome of'How much more important is one species (example) than another." Any pairwise comparison urgently asks the person to accept the conclusion to answer the question: "How much characteristic A can be more important, than characteristic B, of a comparatively common goal?" (Kahraman 2008).

ELimination Et Choice Translating Reality (ELECTRE) is another MCDM technique. The fundamental concept of the ELECTRE method is how to overcome with a leading relationship, using paired comparisons between candidates for any aspect individually. Differences in the two or many choices, significant as $Ai \acute{I} Aj$, indicate that the 2 candidates i and j do not mathematically prevail over each other, the person accepting the conclusion perceives the risk of considering Ai as better than Aj. A candidacy is considered to be dominant if another candidacy overtakes it, at least in 1 aspect, and is equated in the remaining aspects. The ELECTRE method of application is a pairwise comparison of choice based on the degree to which the evaluation of alternatives and the authority of preference recognizes or contradict pair matching with the presence of a predominance between candidates. The decision-maker has the opportunity to say, in fact, that he/she has a strong, weak or indifferent predilection, or even has the ability to be unable to express his preference between the 2 compared candidates (Kahraman 2008).

In comparison to other MCDM methods, PROMETHEE is an efficient technique that provides more preference functions to decision makers for creating the priority to alternatives based on each criteria. The advantages of PROMETHEE include that it is a user-friendly method that can be perfectly applied to real-life problem structures. Both PROMETHEE I and II as whole enable the ranking of the alternatives respectively, while still providing simplicity (Ozsahin et al. 2019).

The PROMETHEE II method arranges objects from the best (more precisely, from the most preferred) to the worst (to the least preferred). To do this, the differences, $Phi = Phi - Phi -$, are calculated for each object and then ordered in descending order. In other words, the ranks of the objects are constructed following the rule: where largest value of Φ is set to a rank equal to 1. As a result, each object receives a rank. The most preferred objects have higher Phi value . In other words, the ranks can be considered as numbers showing ranking of the objects from best to worst (Ozsahin et al. 2019).

References

Balioti V, Tzimopoulos C, Evangelides C (2018) Multi-criteria decision making using TOPSIS method under fuzzy environment. Appl Spillway Select Proc 2(11):637. https://doi.org/10.3390/proceedings2110637

Bellman RE, Zadeh LA (1970) Decision making in a fuzzy environment. Manag Sci 17:141–164

Kahraman C (2008) Multi-criteria decision making methods and fuzzy sets. Springer Optim Its Appl Fuzzy Multi-Criteria Decision Making pp 1–18. https://doi.org/10.1007/978-0-387-76813-7_1

Lee PT, Yang Z (2017) Multi-criteria decision making in maritime studies and logistics: applications and cases. Springer International Publishing, Cham

Ozsahin DU et al (2019) Evaluation and simulation of colon cancer treatment techniques with fuzzy PROMETHEE. In: 2019 Advances in science and engineering technology international conferences (ASET) 2019. https://doi.org/10.1109/icaset.2019.8714509

Ozsahin I, Sharif T, Ozsahin DU, Uzun B (2019) Evaluation of solid-state detectors in medical imaging with fuzzy PROMETHEE. J Instrum 14(01). https://doi.org/10.1088/1748-0221/14/01/c01019

Zadeh LA (1965) Fuzzy sets. Inf Control 8:338–353

Chapter 2
Theoretical Aspects of Multi-criteria Decision-Making (MCDM) Methods

Berna Uzun, Dilber Uzun Ozsahin, and Basil Duwa

Abstract Multi-criteria decision making is recorded as one among many thriving disciplines associate to settling thoughts and issues in relation to the multiple features of the alternatives. Decision is made in daily bases which is part of life. However, it could be associated to a particular person's interest. In other words, decision is made during an intention to either do something or not. These decisions can be considered on intent on what to eat, wear or purchase or a career to choose as the case may be. However, this research centers on making a choice on having a specific character of interest. Many decision making issues have real contradictory objectives in life that are highly considered. Multi-criteria decision-making is considered among many fields that allow selection to take place. Material selection is one of the biggest features to consider in engineering and research. Furthermore, selecting a material to be used in a research or in decision-making area is a distinctive building technique that is important in solving selection problems. Technically, choosing a material or in other times replacing existing material may be as a result of in effectiveness of the first which can easily be replaced by the later. Multi-criteria decision making in other terms can be a qualitative and quantitative analysis. It is applicable incredibly in different areas (fields) of specialization. This work introduces multi-criteria decision

B. Uzun (✉) · D. Uzun Ozsahin
DESAM Institute, Near East University, Nicosia, Turkish Republic of Northern Cyprus, Turkey
e-mail: berna.uzun@neu.edu.tr

D. Uzun Ozsahin
e-mail: dilber.uzunozsahin@neu.edu.tr

B. Uzun
Department of Mathematics, Near East University, Nicosia, Turkish Republic of Northern Cyprus, Turkey

D. Uzun Ozsahin · B. Duwa
Department of Biomedical Engineering, Near East University, Nicosia, Turkish Republic of Northern Cyprus, Turkey

D. Uzun Ozsahin
Medical Diagnostic Imaging Department, College of Health Science, University of Sharjah, Sharjah, United Arab Emirates

D. Uzun Ozsahin et al. (eds.), *Application of Multi-Criteria Decision Analysis in Environmental and Civil Engineering*, Professional Practice in Earth Sciences,
https://doi.org/10.1007/978-3-030-64765-0_2

making approaches in solving and analyzing the problem that tilts towards solving environmental engineering problems, understanding its strengths and weaknesses involved.

Keywords Decision making · Multi criteria decision making process · Multi criteria decision-making techniques

2.1 Introduction to Multi-Criteria Decision-Making Analysis

Knowledge on multi-criteria decision-making (MCDM) can predate to the existence of man. Analytical reasoning and approach is a feature possess by both humans and animals. This, distinctive analytical knowledge gives an individual the potentials to rationally perceive a problem and solves it critically with intelligence. However, MCDM is known since the existence of man without proper documentation and reference.

A renowned American polymath and scientist, Benjamin Franklin (1706–1790), analysed critically on his logical and intellectual perception of using two sides of opinions, contradicting them (Zionts 1979). In other words, Benjamin Franklin argued on and against an opinion, then analyzed them both with a powerful demonstration. He analyzed arguments that are of similar importance, writing them on a paper. After he made an equal balance, it was observed that one side has tremendous supportive argument remaining, which he considered as his decision. Benjamin Franklin design of knowledge earned him a heroic place in drafting of the U.S declaration of Independence and the constitution and other logical negotiation such as that of treaty of Paris.

In a similar work to Benjamin Franklin, Kuhn and Tucker in 1951 defined the MCDM problem using the nonlinear programming condition to optimize the MCDA problems while considering the criteria simultaneously (Charnes and Cooper 1961). Other scientists such as Charnes, Cooper and Fergusson in 1955 established a brandname tagged "Goal Programming" which was later published in 1961 by Cooper and Charnes, respectively (Charnes et al. 1978). Their work, attracted numerous writers and scientists because of the relevance of the work across different fields. These fields include operations research and management sciences. Many researchers and publishers became interested in Cooper and Charnes and contributed immensely to their work, among which include Stan Zionts and Bruno Contini. These individuals mutually worked with cooper to develop and publish a model in 1968, known as the "Multi-Criteria Negotiating Model".

Zionts one of the researchers continued his worked in relation to his previous work at Brussels, European Institute for Advanced Studies in Management and met another fellow named Jyrki Wallenius in 1973. These duos worked amicably using the "goal programming" and developed a "Zionts Wallanius" communicating procedures to solve many linear programming lapses. Subsequently the duos were also joined by another researcher named Pekka Korhonen in 1070. These individuals worked tremendously in putting decision supportive methods used for interaction solving mathematical programming problems. Their method of work attracted many researchers globally. This brought some researchers into limelight; these individuals include Carlos Romera, James Ignizio and Sang Moon Lee as major contributors in goal programming.

In 1959 Ron Howard collaborated with Kimball G. E in 1959 to write an article on the "Sequential Decision Processes". However, the term "Decision Analysis" was considered used first by him in the mid 1960s.

Another historical event is that of Ralph Keeney and Raiffa Howard in 1976 co-authored and published a book that was incredibly important in the Multi Attribute Value Theory establishment. This work is regarded as a standard work that can be of a great reference for generations with regard to the study multi-criteria decision making (MCDM). Subsequently, ELECTRE was developed in Europe by Bernard Roy and others in the mid 1960s, respectively. This idea was to construct a great network of preferences to establish methods that are outstanding. The late Amos Tuersky and Daniel Kahneman worked on behavioral decision theory, which earned Nobel prize in Economics in the year 2002.

2.1.1 Multi-criteria Decision-Making Meetings

Numerous dialogue was met that led to an organized meeting in 1975 by Zionts and in 1977 by Buffalo in Jouy-en-Josas, with other relating researchers such as Fandel Gunter, Tumas Gal, Stan Zionts, Andzej Wierzbicki and Jaap Spronk. These individuals attended a meeting in Konigswinter, Germany related in 1979 that led to founding of Special Interest Group (SIG) on MCDM. This gave Zionts a portfolio of becoming the group leader. These individuals considered some reputable conference, recorded in France, New York and Jouy-en-Josas with interesting packages (founding) attached to them, respectively.

In 1980, J. Morse organized a MCDM conference recorded in Dalaware as the fourth conference recorded and the P. Hansen organized the fifth conference in Mons Belgium in 1982. These meetings were held in different locations around the globe every two years. Yacov Haimes organized the sixth meeting in 1984 in Cleveland Ohio while Y. Sawaragi and H. nakayama organized the seventh conference in 1986 in Japan. A. G Lockett and G. Islei organized the eighth conference in 1988 in Manchester, United Kingdom.

The ninth International conference was organized in 1990 by Ambrose Goicoechea in Fairfax, Virginia. The tenth conference was organized by Gwo-Hshiung and P.L. Yu in 1992 in Taiwa, Taipei province which was hugely assisted by the Taiwanese government; these recorded high profiling individuals such as the Russian Billionaire Boris Berezovsky in attendance. The eleventh conference was in Coimbra (Portugal) in 1994 organized by J. Climaco, while in 1995 the twelfth conference was organized by G. Fandel and T. Gal in Hagen, Germany. The thirteenth conference was organized in Cape Town (South Africa) in 1997 by T. Stewart while the fourteenth conference, organized by Y.Y Haimes in 1998 in Charlottesvile (U.S.A).

The fifteenth conference was organized in Ankara, Turkey in 2000 by M. Kksalan. This was subsequently followed by the sixteenth conference in 2002, organized by M. Luptacik and R. Vetchera in Semmering (Austria) which was followed by the seventeenth conference organized by W. Wedley in 2004 in British Columbia, Canada. The eighteenth conference was organized in Chania (Greece) in 2006 by C. Zopounidis and followed by the Nineteenth conference organized by M. Ehrgott in 2008 in Auckland (New Zealand). The twentieth conference was organized by Y. Shi and S. Wang in June 2009 in Chengdu (China).

These conferences were active and are well organized. This moves simultaneously to the 25th conference that was organized in 2019 in Istanbul, Turkey, which will be followed, subsequently by the June 2021 conference scheduled to take place in Portsmouth, UK.

2.2   The Main Definitions

This study presents an elaborate knowledge on the MCDM, this part also, explains every term introduced in this book of decision analysis involved.

Multiple (Multi):

As the name implies, multiple, is perceived to be diverse in its look. This in other words means is when things are numerous in terms, many or tremendously big.

Criteria:

Etiologically, criteria can be affiliated to criterion in its plural form. This term is perceived to be a form of character or feature possess by an object. This may extend to give a perfect description of anything.

Decision:

Derived from a latin word, means "to cut off". In other words, it is an act of decision that is essential in "cutting off" of anything. Choices are made by individuals or groups about almost everything. It is a mind resolution to accept or reject after tremendous analysis to consider.

Analysis:

This is process of breaking any difficult topic or matter into its smaller forms for a simple and clear understanding. In other words, it can be perceived as a detailed knowledge to examine elements or structure of anything.

Multi Criteria Decision Making (MCDM):

This is a field that deals with decision that hugely involves a better or an outstanding choice from different replicates of ideas based on criteria or attributes that may not be clear to the observer.

Decision Analysis:

Another important aspect of learning is the decision analysis; this in other terms can be a systematic and qualitative visualized approach that tends to evaluate important choices perceived by the decision maker. This can be used by groups and individuals that tend to make decisions about investment, business decisions strategy and even risk management.

Multi Criteria Decision Analysis (MCDA):

MCDA involves solving decision problems systematically. It is considered to be theoretical in its approach using diverse choices. Decision making is considered a field that relates to other fields effectively. An instance, is having an intention of purchasing a vehicle; most importantly the buyer decides on the vehicles efficiency, what to consider when buying the vehicle. However, similar comparable characters evolve whenever an individual decides to buy, eat or even travel.

There are two main important facts that grease our potentials in achieving our goals as individuals. These include;

1. Decision models or theories can be used to assist other individuals, societies or groups to understand the power of decisions and principles attached to any decision taken.
2. Decision models are used to design blue print on any decision to take as a plan for a successful work.

MCDA is regarded as one among many types of analysis that is considered to possess advantageous character towards broad spectrum importance. Choice to do anything is one of human character which extends to solving major problems faced by human when making decisions in all facets of life; career or not. Many alternatives can be evaluated when taking a decision; however, this can be difficult times which may not be satisfactory. In other words, decision makers are engaged to logically employ their knowledge into rationale analysis. This alternatively brings MCDA into limelight used as an appropriate tool by experts to inform, analyze, clarify and justify decisions to be certified with success. This process was used in 1979 applying the Stanley Zionts publication (If not A Rom, Numeral, Then What?) which he tried to persuade his readers to use the method. This method was subsequently adopted in other journals of International Society of MCDM.

2.3   Important Process to Follow

MCDM has essential steps to consider. The analytical decisions highly considered include the following;

STEP 1   Defining The Problem:

This is one of the most important steps to consider when making a decision. Without identifying the problem, the whole system becomes vague. Furthermore, in any case, the essential importance to note is to understand a problem before finding the solution. In many cases people tend to take decisions without understanding them. This could relatively affect the whole decision without solution.

STEP 2   Determination of the Goal:

This process is done after problem is identified with a defined objective to propose the decision making. The goal of any decision making is guided by the end result. In other words, if there is no clear goal for the MCDM, then the result may be affected. For instance, when an individual intends to purchase an object, the person may consider an affordable yet qualitative product, which is regarded here as the objective.

STEP 3   Specify Criteria:

Another important thing to consider is selection of the right criteria in MCDA. This is considered as an attribute that guides an individual in getting the right objective. The criteria selected determine the success of our objectives. However, it is important to consider meaningful criteria when setting an objective. This will enable a better comparison among the options at hand. Criteria make the objective to have meaning. It can be expressed when we compare two individuals. For instance, a person buys a mobile phone and another person buys fruits. These individuals clearly have the same objective even when they bought different items. Their objectives were to buy cheap and qualitative objects. Also, their criteria here are obviously different because they have two different subjects bought. For the individuals that bought fruit, the criteria attached to the purchase of fruit might be quality and durability while the other individual that purchased mobile phone might be quality and its capacity such as camera quality or its version.

STEP 4   Determining the Avaliable Alternatives:

Lists of options are important to design after the criteria are extracted. In other words, getting the right criteria does not complete our determined objective of getting something done, it also enables progressive work. Drawing a list of options will enable our criteria to make more sense for a proper analysis. Another perspective is when we have items that share the same function and quantity but have different cost and quality. This can be seen when applying universities. Sometimes, individuals tend to consider ranking, accommodation and teaching factors as options to consider before applying. Having a designed criteria assist to match our drawn criteria.

STEP 5   Analyzing the Alternatives:

After getting the relative options in line with our objectives, it is now important to

evaluate our options simultaneously. This will assist us to consider consequences associated with our options. The result obtained can assist our logical conclusion on each evaluation. It is however advisable to always consider option looking critically into our objectives so as to have a potential gain with less risks involved.

STEP 6   Calculations:

This step introduces us to the use of data collected as mentioned to calculate and select the best score to consider. This step enables us to obtain result by selecting and picking the product of score for each criterion, getting the weight and then sum the scores together. We obtain the final score when we add all the scores together choosing the option with the best scores.

STEP 7   Reporting the Results:

This process tends to document every detailed desirable results obtained from the previous options. This process assists in preserving all this as a document for future use.

2.4   The strength of the usage of the MCDA

MCDA gives us various processes that are advantageous when compared to other decision-making tools that are not affiliated to any specific criteria, these include the following:

- It is not vague
- It can be adopted for different scenario simply
- It is rational
- It can be applied in various area
- MCDA assist in making decision
- Data combination could ease the decision maker work

Many theories can be applied to this study which can also be summarized showing their advantages.

"Data Envelopment Analysis" (DEA) has the advantage of analyzing numerous input and output with considering and measuring the efficiency togerher (Charnes et al. 1978). In other words, DEA is applicable to many fields such as, economics, medicine, software engineering, road safety, utilities, computer technology, agriculture and solving many business problems. DEA is considerably favorable in resolving problems giving a precise output.

One important, simple and easy method to be considered is the TOPSIS method (Yahya et al. 2020). This is applicable to fields such as technology, transport and economics (Behzadian et al. 2012). However, it can be difficult to weight the importance levels of the criteria.

Another important method is the ELECTRE method. This method is useful because of its ability to accept anything vague into account (Konidari and Mavradis 2007) and applicable to many fields such as transport and water management. It also has disadvantage of its ability to process its end products, which may not be able to be read in simply. This is a difficult process to explain in simple terms.

The PROMETHEE technique is one method that is perceived to be an easy technique that won't need proportionate criteria in assumption (Ozsahin et al. 2018). It is also applicable to many disciplines such as environment, energy, water management, agriculture, education, business, finance and healthcare (Behzadian et al. 2010).

A similar to the PROMETHEE method, Analytic Hierarchy Process (AHP) is also a technique based on pairwise comparison without demanding the exact data. Its ranking structure can also be adjusted to measure different problems involved. This method is also applicable to many fields such as public policy, political strategy, and planning and resource management (Lai 1995). Thus, it has its disadvantages attached such as the interdependence that exists between the options and criteria.

The "Simple Additive Weighting" (SAW) is another method that has the capacity to adapt amid criteria, its perception in making decisions. This method spreads across different fields of study such as in business, finance and water management (Podvezko 2011). However, the method needs significant work to prepare considerate data before it is executed.

The VIKOR method is an MCDA technique that resolves decision-making problems that seem to contradict other problems. In this method, the person making the decision tends to seek the best solution that is ideal to defined criteria while considering the minimum regret (Yu 1973). The method is applicable to engineering.

The Fuzzy Logic Or Fuzzy Set Theory method involves using vague inputs or insufficient data (Zadeh 1965). It extends its application to different fields such as in engineering, economics, social, environmental and business problems involved. One of its disadvantages is that it cannot be developed easily. It needs the experts opinion. Its hybrid application with the other MCDA techniques gives the best solution of the alternatives while the vague data arise.

References

Behzadian M, Kazemzadeh RB, Albadvi A, Aghdasi M (2010) PROMETHEE: a comprehensive literature review on methodologies and applications. Eur J Oper Res

Behzadian M, Otaghsara S, Yazdani M, Ignatius J (2012) A state-of-the-art survey of TOPSIS applications. Expert Syst Appl 39(17):13051–13069

Charnes A, Cooper WW (1961) Management models and industrial applications of linear programming. Wiley, New York

Charnes A, Cooper WW, Rhodes E (1978) Measuring the efficiency of decision making units. EJOR 2:429–444

Konidari P, Mavrakis D (2007) A multi-criteria evaluation method for climate change mitigation policy instruments. Energy Policy 35(12):6235–6257

Lai S (1995) Preference-based interpretation of AHP. Int J Manage Sci 23(4):453–462

Ozsahin I, Uzun B, Isa NA, Mok GSP, Uzun Ozsahin D (2018) Comparative analysis of the common scintillation crystals used in nuclear medicine imaging devices. In: 2018 IEEE nuclear science symposium and medical imaging conference proceedings (NSS/MIC), Sydney, Australia, 2018, pp 1–4. http://doi.org/10.1109/NSSMIC.2018.8824485

Podvezko V (2011) The comparative analysis of MCDA methods SAW and COPRAS. Inzinerine-Ekonomika-Eng Econ 22(2):134–146

Yahya M, Gökçekuş H, Ozsahin D, Uzun B (2020) Evaluation of wastewater treatment technologies using TOPSIS. Desalin Water Treat 177:416–422. https://doi.org/10.5004/dwt.2020.25172

Yu PL (1973) A class of solutions for group decision problems. Manage Sci 19(8):936–946

Zadeh L (1965) Fuzzy sets. Inf Control 8(3):338–353

Zionts S (1979) MCDM—if not a roman numeral, then what?. Interfaces 9(4):94–101. https://www.mcdmsociety.org

Chapter 3
Analytical Hierarchy Process (AHP)

Dilber Uzun Ozsahin, Mennatullah Ahmed, and Berna Uzun

Abstract This study provides a comprehensive explanation about one of the important multi-criteria decision-making technique entitled Analytical Hierarchy Process (AHP). This chapter will include a summary of the steps required to carry out the mathematical computation of AHP for problems with both consistency and inconsistency in the decision-maker's preferences. Hierarchy design, prioritization, criteria weights, and consistency are all extensively taken into consideration. Further elaborations regarding the different applications of the AHP process will also be discussed, including various fields such as business-related decision making and decision theory in the field of medicine. The limitations imposed by the analytical hierarchy process are also discussed.

Keywords Analytical Hierarchy Process · Criteria Weights · Consistency Index · Prioritization · Pair-wise Comparison · Ratio Scales · Eigen Vectors · Eigen Value

D. Uzun Ozsahin · B. Uzun (✉)
DESAM Institute, Near East University, Nicosia, Turkish Republic of Northern Cyprus, Turkey
e-mail: berna.uzun@neu.edu.tr

D. Uzun Ozsahin
e-mail: dilber.uzunozsahin@neu.edu.tr

D. Uzun Ozsahin · M. Ahmed
Department of Biomedical Engineering, Near East University, Nicosia, Turkish Republic of Northern Cyprus, Turkey

B. Uzun
Department of Mathematics, Near East University, Nicosia, Turkish Republic of Northern Cyprus, Turkey

D. Uzun Ozsahin
Medical Diagnostic Imaging Department, College of Health Science, University of Sharjah, Sharjah, United Arab Emirates

D. Uzun Ozsahin et al. (eds.), *Application of Multi-Criteria Decision Analysis in Environmental and Civil Engineering*, Professional Practice in Earth Sciences,
https://doi.org/10.1007/978-3-030-64765-0_3

17

3.1 Introduction

Ultimately, everyone is considered as a decision maker as we are all faced with infinite choices on a daily basis and our decisions are made as result of a certain decision-making process whether it be consciously or unconsciously. We accumulate a certain degree of knowledge to help us reach our decisions. It is instinctive to attempt to gather as much data as possible to reach a sound decision. However, not all the data is valuable for the comprehension and development of judgments and can instead sometimes cause complications and confusion. In some instances, it has been demonstrated that knowing more does not necessarily lead to a better understanding. To fundamentally comprehend a problem, it is necessary to determine the purpose of the decision as well as the objectives and aims as the main factors that are the most influential in choosing between different options. To identify the optimal alternative, prioritization is a key notion in the process (Saaty 2008).

Decision making is a multi-disciplinary process that is a crucial aspect of different processes in many different fields. Thus, attempts have been made for many years to find an automated set of rules based on which decisions can be simplified and clearly made. Using logic, psychology and mathematical computation decision making techniques have been developed and applied in a variety of fields to help automate the process and give consistency to final decisions made by large corporations and systems.

The Analytical Hierarchy Process (AHP) is a multi-criteria decision making technique developed by Saaty (1990) in the 1970s and has been continually advanced and improved upon since then (Saaty 1980). This technique uses a hierarchy framework or levels and then uses pair-wise comparisons to eliminate alternative options; this method is solely based on the comparison of two alternative elements based on a common property. In this way, complex problems can be unravelled and a clear decision can then be made.

To generalize, the analytical hierarchy process is a widely inclusive system used to mathematically compute quantitative and qualitative criteria to help allude to the most appropriate and ideal solution. It is a simple but extremely powerful tool for subjective problems regarding prioritization and selection and is therefore widely applied in multiple fields and organizations. For example, businesses can apply AHP for strategic planning, equipment choices, customer selection and project management, while technological industries can also implement AHP for innovations with several solutions AHP can also be implemented in government, healthcare and even personal decisions such as college choices, property purchases or even when it comes to purchasing a new car or mobile phone. This system can take a subjective problem and give precise discrete nominal quantification values of priorities in a way that is efficient and reliable.

The analytical hierarchy process has been described based on seven foundations (Saaty 2001). The first of these is the derivation of ratio scales from reciprocal pair wise comparisons. These ratio scales represent proportionality and are a crucial component for synthesizing priorities in decision theory. Another main aspect of AHP

Fig. 3.1 Hierarchy
framework

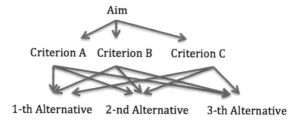

is the psychological origin of the scale used to make said comparisons, which leads to the third pillar which is the inconsistency also denoted as the sensitivity to changes in judgements. Prioritization is then developed giving rise to an eigenvector that can be integrated in the AHP general feedback structure, hence reducing the mathematical method to a one-dimensional normalized ratio scale allowing a unit-less scale measurement. This methodology can be then marked on the basis of whether or not the ranking is preserved or reversal is allowed. Finally, implementing mathematical methodologies into group decision making to generate individual opinions or judgements is crucial to enable the fabrication of a fundamental group decision that would be compatible with individual predispositions.

There are several steps involved in AHP; when given a complex problem, the first step is to construct the hierarchy framework, also known as a feedback network. This hierarchy will have three levels: level one being the goals, level two being the different criteria and finally, level three being the different alternative options as shown in Fig. 3.1. The decision-making process is then initiated by systematically comparing elements two at a time, and giving numerical weight values using scale ratios to each element based on either actual data or subjective opinion. This is where inconsistency can be possible, thus leading to the need to calculate the consistency index, which is a numerical value representing variance or inconsistency. The weights are then used to make a pairwise comparison matrix which is then used to calculate criteria weights. The mathematical computations involved in the AHP are categorized as eigenvector calculations where the Eigen value corresponds to each criteria weight. Depending on the context of the problem being solved, the same could then be applied to sub-criteria if they are present. In this way, we can compare criteria with differing scales, i.e. price and size.

While the analytical hierarchy process is widely applicable to various types of problems that vary in terms of their degrees of complexity, this method is most efficiently applied in more complicated problems that have a large number of objectives or criteria that may include sub-divisions within them (sub-criteria). As such, for more complex problems, it is harder to compare all the objectives to one another and it can be significantly error prone if not approached in a systematic and consistent manner. To resolve this problem, the AHP can be applied to complicated issues such as the prediction of the future of higher education, design choices for a national transport system and even recruitment options in the workplace. In addition to this, AHP provides the decision-maker with quantified measurements of compatibility that can then be utilized to analyse the problem manually (Lee et al. 2007).

In this chapter, we will dissect the mathematical computation of AHP with two simplified examples to help clarify the methodology. The Eigenvalue method is used is used for decision makers that have consistency in their preferences. Consistency is a pivotal concept in AHP as it determines the method to be used. In the case of an inconsistent decision maker, the Eigenvalue method is inapplicable and so the quantification is made through a matrix solution (2). The identification of inconsistency as opposed to consistency in decision making will also be extensively discussed. For example, how can we tell if the decision maker is consistent with their preferences? This will be discussed thoroughly and elaborated with worked examples giving relevance to the distributive and ideal modes of AHP (Saaty 2001).

3.2 Mathematical Computation of AHP

As previously mentioned, the first step in AHP is setting clear objectives and goals. The different criteria are considered along with their sub-criteria and are combined with the alternatives to form the hierarchy tree network. Then, the different qualities of each alternative, the criteria, are assessed in a pairwise comparison with respect to the set objectives to derive their priority as a numerical value using a ratio. The ratio scale or scale of relative importance developed by Saaty (1987) ranges from 1–9, where 1 denotes equal importance and 9 denotes extreme importance, as illustrated in Table 3.1 (Taherdoost 2020). This numerical value is denoted as the criteria weights. These weights are then put into matrix form and mathematical steps are carried out to evaluate the alternative weights and consistency ratios. Therefore, we initially have an input as actual measurements of subjective opinions and our end result will be the ratio scales, which will be in Eigenvector form (denoted as ω),

Table 3.1 Ratio scale of relative importance (Saaty 1987)

Importance	Definition	Explanation
1	Similarly important	Both of the components have the same commitment within the objective
3	Moderately important	One component has a normal advantage compared to the other element
5	Strong important	Having a compelling favouring of one component compared to the other
7	Very solid and demonstrated importance	One element is escalation favoured and has upper control in practice, compared to the other present components
9	Extreme importance	One component is advocated in comparison with the other, this is based on the intensity of the demonstrated evidence and facts
2, 4, 6, 8	Inter-values	

as well as the consistency index, which will be in Eigenvalue form (denoted as λ). The mathematical methodology used here is based on the Eigenvalue problem. The Eigenvector of alternative weights as well as the consistency index is then utilized to rank the alternatives, thus clarifying the optimal alternative.

When faced with a choice between x number of choices, one can apply the AHP method to select the optimal choice. To do this, the decision maker must first design the hierarchy model. This requires an assessment of the problem at hand by categorizing the alternatives together, the criteria and finally, setting a goal. They are then put into a hierarchy network similar to that shown in Fig. 3.1. Secondly, the decision-maker will use pairwise comparisons to compare each choice with the others using the ratio scale of relative importance illustrated in Table 3.1. These comparison values will be used to construct the pairwise comparison matrix A.

$$A = \left[a_{ij} \right]$$

This matrix will have the following characteristics:

- Square matrix with dimensions of $x \times x$, where x represents the number of choices (criteria)
- Leading diagonal input elements will all hold a value of 1.
- Positive matrix; meaning all elements will be greater than 0, non-negative.

$$a_{ij} > 0 \quad \text{for } i, j = 1, \ldots, n$$

- Reciprocal matrix; meaning adjacent inputs will be the reciprocal values of each other.

$$a_{ij} = \frac{1}{a_{ji}} \quad \text{for } i, j = 1, \ldots, n$$

Consistency in judgments is a mandatory concept in AHP that will be thoroughly discussed in this chapter. To generalize, it legitimizes the applicability of the AHP method depending on how consistent the decision-maker is when applying his/her personally motivated preference in judgements. If the following holds true, then the decision maker is said to be consistent:

$$a_{ik} = a_{ij}^* a_{kj} \quad \text{for } i, j, k = 1, \ldots, n$$

In AHP, inconsistency is expected and accounted for. This is because the numerical values are derived from the decision maker's preference or individual opinions. In real life, these values can be inconsistent and therefore these inconsistencies must be accounted for. To summarize, below is a step-by-step simplified overview of the methodology followed by several simple examples that will be used to further reinforce a more developed understanding (Saaty 2008).

1. The decision maker must define the problem and determine the kind of data they will make the judgments on.
2. The hierarchy is structured and designed, designating the top level as the defined goal, the second level as the broader interpretation of the determined objectives, intermediate levels as the sub-criteria to be assessed and finally, the lowest level will be composed of the different alternative options.
3. Using the ratio scale of relative importance developed by Saaty 1987 (Table 1.1), a pairwise comparison matrix is constructed where all the elements are comparative judgements made by the decision maker.
4. The pairwise matrix is normalized, and the criteria weights are derived as the average of each row in the matrix.
5. Using the derived criteria weights, we can find the Eigen vector (ω) by calculating the weight sum value for each criterion.
6. The Eigen value (λ) is then found such that $A\omega = \lambda\omega$
7. Finally, the consistency index is calculated to legitimize the reliability of the decision maker's judgments.

3.3 Theoretical Apprehension of Consistency

The concept of consistency is one of the essential step of the analytical hierarchy process. In principle, the consistency ratio is calculated as the reliability of the preferential judgments in comparison to a large number of randomly generated judgments. Realistically, inconsistency is inevitable as it is non-avoidable, primarily because the foundation of decision making is based on the personal preference of the decision maker and it is inevitable that inconsistency will occur in the preference of the decision maker. In other words, the input of the AHP system is based on personal preference and therefore highly prone to human error (Dyer and Forman 1991). The question at hand is the degree of the consistency and whether or not it satisfies the predetermined standard values (Mu and Pereyra-Rojas 2017).

To quantify the level of consistency in the problem at hand, the consistency ratio is derived as the ratio of the consistency index to the random index. The consistency index represents the consistency of the pairwise matrix of the given problem. On the other hand, the random index matrix is a representation of the average consistency ratio of 500 randomly generated pairwise matrices. These values are predetermined and constant; they are singularly dependent on the dimensions (n) of the problem at hand. The calculated values are illustrated in Table 3.2. Based on the previous work of Saaty regarding the complexities of the concept of consistency in the analytical hierarchy process, the consistency ratio is defined as CR where CR = CI/RI (Saaty 2012). He also eluded that the standard CR value is 0.1, meaning that if the consistency ratio is calculated to be equivalent to or less than the standard 0.1 value, then the problem is acceptably denoted as consistent and the analysis process is legitimized.

Table 3.2 Corresponding random index values for different matrix dimensions

Matrix dimensions (n)	Random Index value (RI)
1	0.0
2	0.00
3	0.58
4	0.90
5	1.12
6	1.24
7	1.32
8	1.41
9	1.45
10	1.49

Otherwise, if the consistency ratio is found to be greater than the standard value of 0.1, then the problem is insufficiently consistent and requires re-evaluation. In the case where the problem is denoted as inconsistent, the decision-makers must re-assess the preferential reasoning and identify the source of the variance or inconsistencies to then rectify and refine them so that consistency holds true for the problem at hand.

Notice that as the number of criteria increases or as the matrix dimension gets larger, a subsequent incremental effect is imposed onto the corresponding index value as well. This indicates that as the problem becomes more complex, meaning it encompasses a greater number of criteria and sub-criteria (dimensions), then the possibility of inconsistencies in judgments also increases. Relating this ideology to logic helps enhance the fundamental understanding of the concept. In this sense, if we increase the instances where the decision-maker has to provide an input based on personal preference, i.e. criteria, then a higher chance of variations will be induced.

The work done by Saaty proved that for the consistency problem, the pairwise matrix is reciprocal and positive and holds a maximum Eigen value equivalent to the dimensions of the comparison matrix (n). In other words:

$$\lambda\max = n$$

In addition to this, he introduced the consistency index as a measure of consistency and this is found using the formula,

$$C.I = \frac{\lambda\max - n}{n - 1}$$

where n is the dimension of the reciprocal positive comparison matrix and λ max represents the maximum Eigen value, which in the ideal case (no inconsistencies) would hold the identical value n (dimension of the matrix). The closer the maximum Eigen value is to the dimension of the matrix, the more the consistency index decreases, and the consistency ratio decreases accordingly.

References

Dyer R, Forman E (1991) An analytic approach to marketing decisions. Prentice-Hall International, London

Lee M, Wang H, Wang H (2007) A method of performance evaluation by using the analytic network process and balanced score car. In: 2007 International conference on convergence information technology (ICCIT 2007), Gyeongju, pp 235–240. https://doi.org/10.1109/iccit.2007.216

Mu E, Pereyra-Rojas M (2017) Practical decision making. Springer International Publishing, Cham

Saaty T (1980) In: The analytic heirarchy process, McGraw-Hill

Saaty R (1987) The analytic hierarchy process—what it is and how it is used. Mathematical Modelling 9(3-5):161–176

Saaty T (1990) In: The analytical heirarchy process: planning proprity setting resource allocation

Saaty T (2001) In: The seven pillars of the analytical hierarchy process. Pittsburgh, PA 15260, USA, University of Pittsburgh

Saaty T (2008) Decision making with the analytic hierarchy process. Int J Serv Sci 1(1):83

Saaty T (2012) Decison making for leaders: the analytical heirarchy process for decisons in a complex world. RWS Publications, Third Revised Edition, Pittsburgh

Taherdoost H (2020) Decision making using the analytical heirarchy process (ahp); a step by step approach. Int J Econom Manage Syst IARS 2:244–246

Chapter 4
The Technique For Order of Preference by Similarity to Ideal Solution (TOPSIS)

Berna Uzun, Mustapha Taiwo, Aizhan Syidanova, and Dilber Uzun Ozsahin

Abstract The Technique for Order of Preference by Similarity to Ideal Solution (TOPSIS) is the part of the analytical multi-criteria decision-making technique. The main idea of this technique, the preferred alternative is the one with the most close to the positive ideal solution and the further to the negative ideal solution. The positive ideal solution is formed as a combination of the best points of each criterion. The negative ideal solution is a combination of the worst points of each criterion. This technique is only can be applied for the numerical dataset where the importance weights of the criterion known or defined based on the experts opinion numerically. And the ranking results can be obtained corresponding the importance weights of the defined criteria. In this study the detailed information about this technique will be presented.

Keywords Decision making · TOPSIS · Multi-criteria decision-making

---

B. Uzun (✉) · D. Uzun Ozsahin
DESAM Institute, Near East University, Nicosia, Turkish Republic of Northern Cyprus, Turkey
e-mail: berna.uzun@neu.edu.tr

D. Uzun Ozsahin
e-mail: dilber.uzunozsahin@neu.edu.tr

B. Uzun
Department of Mathematics, Near East University, Nicosia, Turkish Republic of Northern Cyprus, Turkey

M. Taiwo · D. Uzun Ozsahin
Department of Biomedical Engineering, Near East University, Nicosia, Turkish Republic of Northern Cyprus, Turkey

A. Syidanova
Department of Architecture, Near East University, Nicosia, Turkish Republic of Northern Cyprus, Turkey
e-mail: aizhansyidanova@gmail.com

D. Uzun Ozsahin
Medical Diagnostic Imaging Department, College of Health Science, University of Sharjah, Sharjah, United Arab Emirates

D. Uzun Ozsahin et al. (eds.), *Application of Multi-Criteria Decision Analysis in Environmental and Civil Engineering*, Professional Practice in Earth Sciences,
https://doi.org/10.1007/978-3-030-64765-0_4

4.1 Introduction

TOPSIS is an acronym for Technique for Order Preference by Similarity to Ideal Solution. Method TOPSIS first introduced by Yoon and Hwang in 1981 (Hwang and Yoon 1981) and improved in 1987 by Yoon (Yoon 1987), after that in 1993 also by Hwang, Lai and Liu (Hwang et al. 1993). Its fundamental concept ought to have the most limited distance from the positive ideal solution (PIS) and must be far from the negative ideal solution (NIS).

Multi criteria decision-making (MCDM) strategies, TOPSIS could be a usefull and valuable strategy for positioning and selecting the best options by measuring Euclidean distances. TOPSIS could be a straightforward positioning strategy and application. The TOPSIS strategy proposed as a decision making model which can be applied in numerous areas. The major idea of this methodology is obtaining the best imagined alternative (PIS) and then discover a situation, which is closest to the PIS and furthest to the NIS (Sianaki 2020).

The PIS is characterized as the combination of all the most excellent values that can be achieved for each trait, whereas the NIS includes all of the most notable awful scores achieved for each criterion. Based on comparisons with relative distance, elective priority arrangement of action can be accomplished. This strategy is broadly utilized to unravel practical decisions (Ic 2012). TOPSIS is broadly utilized for reasons such as:

- The concept is straightforward and simple to get.
- Having the capacity to measure the relative execution of choice options in a basic numerical frame.

The strategy of TOPSIS based on the a general notion that the leading chosen elective doesn't just have the most limited range from the PIS but moreover has the longest range from the NIS. The steps involved in TOPSIS include:

1- Constructing the decision matrix and importance weights of the criteria based on the decision maker preferences
2- Counting the normalized decision matrix
3- Counting the weighted normalized decision matrix
4- Obtaining a PIS and NIS
5- Counting the separation measures from the PIS and NIS
6- Counting the relative closeness to the PIS

Advantages of the TOPSIS include;

- The concept is straightforward, simple to obtain and able to evaluate a single alternative.
- The concept is comprehensible.
- Efficient computation.
- The differences between the alternatives can be visualized using normalized values (Kraujalienė 2019).
- Instinctive and rational logic that forms the basis of human choice.

However, the shortcoming of TOPSIS includes;

- There might be a weights calculated utilizing AHP or fuzzy logic and later TOPSIS.
- It is difficult to weight as well as keeping the consistency of judgment.
- Euclidean distance's application does not correlate with attributes (Robbi Rahim 2018).
- It is highly subjective (Sharma et al. 2020)

In TOPSIS there is no any capacity restriction, which ease the decision maker work. It is applicable where the huge number of the criteria and alternatives arises specifically if there is objective or quantitative information available for the case.

Generally, the TOPSIS methods algorithms begin with obtaining the decision matrix which represents the combination of the criteria of each alternative. Also, the framework is normalized with one of the normalization technique, and the normalized values are multiplied by the importance weights of their corresponding criteria. Along these lines, the PIS and NIS are arranged, then separate measures of each alternative to these arrangements are calculated based on a distance degree. Lastly, the choices are positioned related to their relative closeness to the PIS. The TOPSIS procedure is accommodating for choice producers to structure the issues to be optimized, conduct investigations, comparisons and positioning of the choices.

The classic method of TOPSIS consists issues where the chosen comparative information are known and presented numerically. In reality, the problems of the world are mostly very complex, which is not easy to model since there is many affect even the decision maker cannot be aware to consider, or suddenly some situations can change, which should be considered in the model to increase its effectiveness. The hybrid applications of the classic TOPSIS method with other extensional models such as; where there is the interval in decision matrix arise or fuzzy criteria while defining the vagueness of the problem, are the new approach that started to be studied by many researcher.

Interval investigation may be a basic and instinctive way to present information, instability for complex choice issues, and can be utilized for numerous viable applications. An expansion of the TOPSIS method to a gather choice environment is additionally explored. The setting of multi-criteria decision-making in vague and interval information could need a hybrid methodology, first the arrangement of the data and then the analysis of the ranking. TOPSIS strategy can only be applied where the decision matrix and importance weights of the criteria supposed to be defined as the numerical data.

4.1.1 Mathematical Framework of the TOPSIS Method

To understand this technique and see how it can be applied, it is important to analyze each step consecutively as shown below (Yahya et al. 2020).

Step 1 Construct a decision matrix and define the importance weights of the criteria.

The decision matrix $X = X_{ij}$ and a weighting vector $W = [w_1, w_2, \ldots, w_n]$ are chosen. Where $X_{ij} \in R$, $W_j \in R$ and $w_1 + + \ldots + w_n = 1$.

Function criteria can either be a "benefit function" (more benefit better alternative) or a "cost function" (low cost better alternative).

Step 2 Obtaining the normalized decision matrix.

In this step, the different kinds of dimensions of the criteria are converted into one dimension that allows comparison between all criteria. To convert all data to normalized form, each element of the matrix X must be converted, since most criteria are usually measured in different units. The normalization of these values can be done using one of the normalization formulas. The normalized values of the decision matrix (n_{ij}) can be obtained with one of the following formulas:

$$n_{ij} = \frac{X_{ij}}{\sqrt{\sum_{i=1}^{m} X_{ij}^2}} \tag{4.1}$$

$$n_{ij} = \frac{X_{ij}}{\max_i X_{ij}} \tag{4.2}$$

$$n_{ij} = \begin{cases} \frac{x_{ij} - \min_i x_{ij}}{\max_i x_{ij} - \min_i x_{ij}} \\ \frac{\max_i x_{ij} - x_{ij}}{\max_i x_{ij} - \min_i x_{ij}} \end{cases} \tag{4.3}$$

Step 3 Obtaining the weighted normalized decision matrix.

Method to obtain the weighted normalized value of the decision matrix (v_{ij});

$$v_{ij} = w_j n_{ij} \tag{4.4}$$

where $i = 1, \ldots, m$; $j = 1, \ldots, n$. And w_j is the weight of the j-th criteria where $\sum_{j=1}^{n} w_j = 1$.

Step 4 Identify the positive ideal and negative ideal solutions.

Here, the PIS on each criterion (the best performance) and the NIS (the worst performance) are identified. The PIS maximizes the benefit criteria and minimizes the cost criteria, while the NIS maximizes the cost criteria and minimizes the benefit criteria.

The PIS A^+ can be obtained with the following formula:

$$A^+ = (v_1^+, v_2^+, \ldots, v_n^+) = \left[\left[\max_i v_{ij} | j \in I \right], \left[\min_i v_{ij} | j \in J \right] \right] \tag{4.5}$$

And oppositely the NIS A^- can be obtained with the following formula:

$$A^- = (v_1^-, v_2^-, \ldots, v_n^-) = \left[\left[\min_i v_{ij} | j \in I \right], \left[\max_i v_{ij} | j \in J \right] \right] \qquad (4.6)$$

where I defined with the benefit criteria and J defined with the cost criteria, $i = 1, \ldots, m$; $j = 1, \ldots, n$.

Step 5 Obtaining the separation values of the alternatives and PIS and separation values of the alternatives and NIS.

In this step several distance metrics can be used. Every alternative from the PIS is separated based on the following formula;

$$d_i^+ = \left(\sum_{j=1}^n (v_{ij} - v_j^+)^p \right)^{1/p}, \quad i = 1, 2, \ldots, m. \qquad (4.7)$$

Every alternative from the NIS is separated based on the following formula;

$$d_i^- = \left(\sum_{j=1}^n (v_{ij} - v_j^-)^p \right)^{1/p}, \quad i = 1, 2, \ldots, m. \qquad (4.8)$$

where $p \geq 1$. The most commonly used distance metric in this step is the traditional n-dimensional Euclidean metric (where p = 2) as given in Formulas 4.9 and 4.10.

$$d_i^+ = \sqrt{\sum_{j=1}^n \left(v_{ij} - v_j^+ \right)^2}, i = 1, 2, \ldots, m, \qquad (4.9)$$

$$d_i^- = \sqrt{\sum_{j=1}^n \left(v_{ij} - v_j^- \right)^2}, i = 1, 2, \ldots, m. \qquad (4.10)$$

Step 6 Calculating the relative closeness to the PIS.

The relative closeness of the i-th alternative A_J concerning A^+ is defined as;

$$R_i = \frac{d_i^-}{d_i^- + d_i^+}, \qquad (4.11)$$

where $0 \leq R_i \leq 1, i = 1, 2, \ldots, m$.

In this technique, the alternate closer to the PIS and the further to the NIS should be preferred (Assari et al. 2012). Corresponding to this argument, the alternatives are ranked based on the closeness to the PIS. The alternative with a higher R_i is the alternative closest to the PIS. The ranking results can be obtained accordingly.

References

Assari A, Mahesh T, Assari E (2012) Role of public participation in sustainability of historical city: usage of TOPSIS method. Indian J Sci Technol 5(3):2289–2294

Hwang CL, Lai YJ, Liu TY (1993) A new approach for multiple objective decision making. Comput Oper Res 20(8):889–899. https://doi.org/10.1016/0305-0548(93)90109-v

Hwang CL, Yoon K (1981) Multiple attribute decision making: methods and applications. Springer, New York

Ic Y (2012) An experimental design approach using TOPSIS method for the selection of computer-integrated manufacturing technologies. Robot Comput-Integr Manuf 28(2):245–256

Kraujalienė L (2019) Comparative analysis of multicriteria decision-making methods evaluating the efficiency of technology transfer. Bus Manage Educ 17:72–93

Robbi Rahim AP (2018) Technique for order of preference by similarity to ideal solution (TOPSIS) method for decision support system in top management. Int J Eng Technol

Sharma D, Sridhar S, Claudio D (2020) Comparison of AHP-TOPSIS and AHP-AHP methods in multi-criteria decision-making problems. Int J Indus Syst Eng 34(2):203. https://doi.org/10.1504/ijise.2020.105291

Sianaki OA (2020, May 29) *TOPSIS:* Technique for order preference by similarity to ideal solution. Retrieved from Maths Works File Exchange: https://www.mathworks.com/matlabcentral/fileexchange/57143-topsis-technique-for-order-preference-by-similarity-to-ideal-solution

Yahya M, Gökçekuş H, Ozsahin D, Uzun B (2020) Evaluation of wastewater treatment technologies using TOPSIS. Desalin Water Treat 177:416–422. https://doi.org/10.5004/dwt.2020.25172

Yoon K (1987) A reconciliation among discrete compromise situations. J Oper Res Soc 38(3):277–286. https://doi.org/10.1057/jors.1987.44

Chapter 5
ELimination Et Choix Traduisant La REalité (ELECTRE)

Berna Uzun, Rwiyereka Angelique Bwiza, and Dilber Uzun Ozsahin

Abstract The "ELimination Et Choix Traduisant la REalité" ELECTRE strategy bring together a subject of decision assistance strategies whose particularity is the partial collection based on the development of relations of comparisons of the exhibitions of each pair of arrangements. It is an outranking strategy dependent on concordance examination. Its significant preferred position is that it is considered vulnerable. One weakness that the procedure also the results could be difficult with clarifying. In other words, it is based on pairwise predominance comparisons between choice focuses for each criterion. ELECTRE has been utilized in different fields such as: financial matters, environment, water management, and transportation problem. In this study, the detailed information of the ELECTRE technique will be presented.

Keywords Decision making · Multi criteria decision making · Concordance set · ELECTRE

B. Uzun (✉) · D. Uzun Ozsahin
DESAM Institute, Near East University, Nicosia, Turkish Republic of Northern Cyprus, Turkey
e-mail: berna.uzun@neu.edu.tr

D. Uzun Ozsahin
e-mail: dilber.uzunozsahin@neu.edu.tr

B. Uzun
Department of Mathematics, Near East University, Nicosia, Turkish Republic of Northern Cyprus, Turkey

R. A. Bwiza · D. Uzun Ozsahin
Department of Biomedical Engineering, Near East University, Nicosia, Turkish Republic of Northern Cyprus, Turkey

D. Uzun Ozsahin
Medical Diagnostic Imaging Department, College of Health Science, University of Sharjah, Sharjah, United Arab Emirates

5.1 Introduction

ELimination Et Choix Traduisant la REalité (ELECTRE) technique is a multi criteria decision analysis, which based on the comparison, by criterion putting forward a preference/indifference of a criteria to another and resulting in an over positioning matrix (Figueira et al. 2005). It has been presented by the Beneyoun within the mid-1960s, and his colleagues at SEMA consultancy company (Roy 1968). Bernard Roy was called in as a expert and bunch devised the ELECTRE strategy. Because it was to begin with connected in 1965, the ELECTRE strategy was to select the leading activity from a given set of activities, but it was soon connected to three fundamental issues: choosing, positioning and sorting. The strategy got to be more broadly known when a paper by B. Roy showed up in a French operation research journal.

The ELECTRE strategy is popular for its outranking relations to rank a set of choices (Dodgson et al. 2020). As an expansion, engineered weight, counting subjective and objective weights, it is developed in result of concordance and no discordance tests including a particular input preference information. This strategy is considered to be generally complex strategy, the most reason being that a number of specialized parameters are considered in this specific strategy additionally the calculation that's embraced is marginally complex in comparison (Fei et al. 2019). These strategy have the advantage of tolerating circumstances of incomparability with subjective and immense criteria.

The method proceeds to the solution in 8 steps as shown below (Alper and Başdar 2017):

Step 1 Obtaining the Decision Matrix (A)

In this step the decision point and the related criteria should be combined in a matrix form. The alternatives should be presented in raw of the decision matrix and the criterion should be presented in column of the decision matrix. For example, if there is m-alternative with n-column, then the decision matrix should be m × n in size as shown below as A_{ij} where $i = 1, 2, \ldots, m$ and $j = 1, 2, \ldots, n$).

$$A_{ij} = \begin{bmatrix} a_{11} & \cdots & a_{1n} \\ \vdots & \ddots & \vdots \\ a_{m1} & \cdots & a_{mn} \end{bmatrix} \tag{5.1}$$

Step 2 Calculating the Standard Decision Matrix (X)

The elements of the Standard Decision Matrix (x_{ij}) can be calculated by the Formula below based on the elements of the decision matrix (a_{ij}).

$$x_{ij} = \frac{a_{ij}}{\sqrt{\sum_{k=1}^{m} a_{kj}^2}} \tag{5.2}$$

And the standard matrix X should be obtained as in the form below:

$$X_{ij} = \begin{bmatrix} x_{11} & \cdots & x_{1n} \\ \vdots & \ddots & \vdots \\ x_{m1} & \cdots & x_{mn} \end{bmatrix} \tag{5.3}$$

Step 3 Calculating the Weighted Standard Decision Matrix (Y)

The significance of assessment factors (criteria) for the selection maker can also additionally vary. In set up to mirror those noteworthiness contrasts inside the ELECTRE arrangement, the Y matrix need to be calculated. In set up to do this, the selection maker need to first of all determine the burden of the evaluation additives (w_i), where:

$$\sum_{i=1}^{n} w_i = 1 \tag{5.4}$$

Then, the Y matrix is made via way of means of multiplying the additives in every column of the X community with the evaluating (w_i) esteem as seemed below:

$$Y_{ij} = \begin{bmatrix} w_1 x_{11} w_2 x_{12} & \cdots & w_n x_{1n} \\ \vdots & \ddots & \vdots \\ w_1 x_{m1} w_2 x_{m2} & \cdots & w_n x_{mn} \end{bmatrix} \tag{5.5}$$

Step 4 Determination of Concordance Sets (C_{kl}) and Discordance Sets (D_{kl})

For the assurance of the concordance units of the options, the Y matrix should be utilized and the choice focuses ought to be as compared with every different in phrases of every assessment calculate. The concordance units can be characterized through the relationship appeared inside the equation below:

$$C_{kl} = \{ j, y_{kj} \geq y_{lj} \} \tag{5.6}$$

This equation fundamentally relies upon on a contrast among the values of the decision matrix row components. In a numerous decision issue with m options, the wide variety of concordance units is (m.m–m). The wide variety of concordance set components maximum extreme can be damage even with to the wide variety of assessment components (n).

As an example, where $k = 1$ and $l = 2$, C_{12} concordance set should be calculated by the comparison of the $1 - th$ and $2 - nd$ rows and for the 5 criteria, C_{12} should consits a maximum of 5 elements. In this example, if the comparison results are as follows:

$$y_{11} > y_{21}$$
$$y_{12} > y_{22}$$
$$y_{13} < y_{23}$$
$$y_{14} = y_{24}$$
$$y_{15} < y_{25}$$
$$C_{23} = \{1, 2, 4\}. \tag{5.7}$$

In the ELECTRE procedure, for every concordance set, a discordance set exists, that is the supplement set of the concordance set. The discordance units include the components j, which aren't contained inside the comparing concordance set.

In the example, due to the fact that $C_{23} = \{1, 2, 4\}$, then $D_{23} = \{3, 5\}$.

When developing the concordance units inside the ELECTRE strategy, the signification and the point of the assessment variables ought to be carefully watched. For example, if the important assessment figure point is maximization, at that point Eq. (5.15) will be utilized for the concordance set. However, in case the point of the assessment is minimization, at that point the circumstance of being inside the concordance units will be inverse as $y_{kj} < y_{lj}$

Step 5 Creation of Concordance (C) and Discordance Matrices (D)

Concordance units are utilized for making the concordance matrix (C). Matrix C is m×m in measure and does now no longer take a esteem wherein k = 1. Components of the lattice C are calculated with the relationship appeared inside the equation below.

Concordance sets are used for creating the concordance matrix (C). Matrix C is mxm in size and does not take a value where $k = l$. Elements of the matrix C are calculated with the relationship shown in the formula below.

For example, if $C_{12} = \{1, 3\}$ the c_{12} element of the matrix C will be calculated as $c_{12} = w_1 + w_3$. The matrix C is shown as below:

$$C = \begin{bmatrix} - & c_{12} & c_{13} & \cdots & c_{1m} \\ c_{21} & - & c_{23} & \cdots & c_{2m} \\ & & \cdot & & \\ & & \cdot & & \\ & & \cdot & & \\ c_{m1} & c_{m2} & c_{m3} & \cdots & - \end{bmatrix} \tag{5.8}$$

The elements of the discordance matrix (D) are calculated by the formula below:

$$d_{kl} = \frac{\max|y_{kj} - y_{lj}|_{j \in D_{kl}}}{\max|y_{kj} - y_{lj}|_{\forall j}} \tag{5.9}$$

For illustration, from the contrast of the components of the first and second traces of the Y matrix, for calculation of the d_{12} (where in $k = 1$ and $l = 2$) wherein the values $j = 2, 4, 5$ ought to be taken into consideration from the components of the discordance set $D_{23} = \{2, 4, 5\}$ and the greatest supreme esteem of $|y_{12} - y_{22}|$, $|y_{14} - y_{24}|$ and $|y_{15} - y_{25}|$ ought to be selected for the nominator and for the denominator of the equation, the maximum extreme supreme distinction of the 1-th and 2-nd line of the framework Y is selected.

Like the C framework, the D framework is $m \times m$ in measure and does now no longer take values wherein $k = 1$. Network D is appeared below:

$$D = \begin{bmatrix} - & d_{12} & d_{13} & \dots & d_{1m} \\ d_{21} & - & d_{23} & \dots & d_{2m} \\ & & \cdot & & \\ & & \cdot & & \\ & & \cdot & & \\ d_{m1} & d_{m2} & d_{m3} & \dots & - \end{bmatrix} \qquad (5.10)$$

Step 6 Creation of the Concordance Superiority (F) and Discordance Superiority (G) Matrices

The concordance superiority matrix (F) is a matrix $m \times m$ in size and the components of this matrix are gotten through evaluating the concordance limit $(\underline{c})$ with the components of the concordance matrix,(c_{kl}), in which the concordance threshold $(\underline{c})$ is obtained with the equation below:

$$\underline{c} = \frac{1}{m(m-1)} \sum_{k=1}^{m} \sum_{l=1}^{m} c_{kl} \qquad (5.11)$$

In the equation, m denotes the number of decision point and the $(\underline{c})$ value is calculated through multiplying the all of the components of the C matrix with $1/m(m^{-1})$. The components (f_{kl}) of the matrix F take an esteem of both 1 or 0 as visible in Eq. (5.12) and in its diagonal there should not be any value due to the fact it appears the identical choice points.

$$f_{kl} = \begin{cases} 1, \; if \; c_{kl} \geq \underline{c} \\ 0, \; if \; c_{kl} < \underline{c} \end{cases} \qquad (5.12)$$

The discordance superiority matrix (G) is a matrix $m \times m$ in size and is obtained in the same way. The value of the discordance threshold value $(\underline{d})$ may be calculated with the following equation.

$$\underline{d} = \frac{1}{m(m-1)} \sum_{k=1}^{m} \sum_{l=1}^{m} d_{kl} \qquad (5.13)$$

In different words, the $(\underline{d})$ value is obtained through multiplying the whole of the components of the D matrix with $1/m(m-1)$.

The components of the matrix G (g_{kl}) moreover take the esteem 1 or 0 as seen in Eq. (5.14), and there is no value on its diagonal as it shows similar decision points. If:

$$g_{kl} = \begin{cases} 1, & if\ d_{kl} \geq \underline{d} \\ 0, & if\ d_{kl} < \underline{d} \end{cases} \tag{5.14}$$

Step 7 Creation of Total Dominance Matrix (E)

Elements of the entire dominance matrix (E) are equal to the cross product of f_{kl} and g_{kl} components as appeared inside the equation underneath. Here, the matrix E is $m \times m$-sized relying on the networks C and D, and it comprises of 1 or 0 values.

Step 8 Determining the Importance Order of Decision Points

The rows and columns of lattice E appear decision points. For illustration, E matrix is calculated as follows,

$$E = \begin{bmatrix} - & 0 & 1 \\ 1 & - & 0 \\ 0 & 0 & - \end{bmatrix} \tag{5.15}$$

$e_{21} = 1, e_{13} = 1$. This appears the supreme predominance of the 2-nd choice over the 1-th choice, and absolute predominance of the 1-th choice over the 3-th choice. In this case, in case A_i ($i = 1, 2, 3$) indicates the decision points, the significance arrange will be decided as A_2, A_1 and A_3.

References

Alper D, Başdar C (2017) A comparison of TOPSIS and ELECTRE methods: an application on the factoring industry. Bus Econ Res J 8(3):627–646

Dodgson JS, Spackman M, Pearman AD, Phillips LD (2020) (PDF) Multi-criteria analysis: a manual [Online] Available at https://www.researchgate.net/publication/30529921_Multi-Criteria_Ana lysis_A_Manual . Accessed 20 October 2020

Fei L, Xia J, Feng Y, Liu L (2019) An ELECTRE-based multiple criteria decision making method for supplier selection using dempster-shafer theory. IEEE Access 7:84701–84716

Figueira J, Greco S, Ehrgott M (2005) Multiple criteria decision analysis: state of the art surveys. New York: Springer Science + Business Media, Inc. ISBN 0-387-23081-5

Roy B (1968) Classement et choix en présence de points de vue multiples. Revue Française D'informatique Et De Recherche Opérationnelle 2(8):57–75. https://doi.org/10.1051/ro/196802 v100571

Chapter 6
Preference Ranking Organization Method for Enrichment Evaluation (Promethee)

Berna Uzun, Abdullah Almasri, and Dilber Uzun Ozsahin

Abstract Many forms of decision aid have been suggested with multi-criteria decision analysis. All of these methods start from the same assessment table, but vary depending on the additional information they request. The Preference Ranking Organization Method For Enrichment Evaluation (PROMETHEE) approaches need very simple external knowledge, which both decision-makers and observers quickly access. The goal of all methods with multi-criteria is to enrich the dominance graph, i.e. reduce the incomparabilities. Once the utility function is designed, the question of multi-criteria is reduced to a single problem of criterion for which there is an optimal solution. It sounds simplistic as it depends on pretty firm premises and totally changes the decision issue structure. This is why Brans suggested building outstanding relationships including only realistic enrichments of the relationship of dominance. In that case, the incomparabilities are not all removed but the information is reliable. In this study the detailed information of the PROMETHEE technique will be presented.

Keywords Decision making · Multi criteria decision making · Preference function · PROMETHEE

B. Uzun (✉) · D. Uzun Ozsahin
DESAM Institute, Near East University, Nicosia, Turkish Republic of Northern Cyprus, Turkey
e-mail: berna.uzun@neu.edu.tr

D. Uzun Ozsahin
e-mail: dilber.uzunozsahin@neu.edu.tr

B. Uzun
Department of Mathematics, Near East University, Nicosia, Turkish Republic of Northern Cyprus, Turkey

A. Almasri · D. Uzun Ozsahin
Department of Biomedical Engineering, Near East University, Nicosia, Turkish Republic of Northern Cyprus, Turkey

D. Uzun Ozsahin
Medical Diagnostic Imaging Department, College of Health Science, University of Sharjah, Sharjah, United Arab Emirates

© The Author(s), under exclusive license to Springer Nature Switzerland AG 2021
D. Uzun Ozsahin et al. (eds.), *Application of Multi-Criteria Decision Analysis in Environmental and Civil Engineering*, Professional Practice in Earth Sciences,
https://doi.org/10.1007/978-3-030-64765-0_6

6.1 Introduction

Brans suggests the PROMETHEE approach and it is one form of method based on an outrank relationship between alternate pairs (Brans 1982). The Promethean and Gaia are best known in their concise additional geometrical analysis for interactive help. Instead of pointing to a correct decision, the Promethean and Gaia approach allows decision makers to find the best answer for their intent and understanding of the problem (Brans and De Smet 2016). It offers an integrated and logical context in which the decision problem can be organized, its contradictions and synergies defined and quantified, action clusters, and the key alternatives and systematic thought underlying them. The Promethean and Gaia will refer to decisions in the following situations like Option (Select one alternative from a specified collection of alternatives, typically when several decision criteria have been included.), Prioritization (Assess the relative value of members of a set of alternatives as compared to the preference of the particular one, or simply identify the alternative.), Allocation of resources to a variety of choices., Classification (Adding a variety of alternatives to make Conflict the most common.) and Resolution (Disagreements between seemingly conflicting parties are settled.

The outranking test compares pairs of alternatives for any parameter. The PROMETHEE approach produces the preferential function to define the disparity in choice on each criterion between alternative pairs (Vincke and Brans 1985). Thus preference functions are built up about the numerical difference between pairs of alternatives to describe the preferential difference from the point of view of the decision maker. The value of such functions varies from 0 to 1. The greater the value of the function, the greater will be the difference of preference. There's no preferential differential between pair of alternatives when the value is zero.

The steps of obtaining the ranking result via PROMETHEE technique is as follows (Ozsahin et al. 2019):

Let $a_1, a_2, \ldots, a_m$ be m alternatives and $q_1, q_2, \ldots, q_n$ be n cardinal criteria and let q_{ij} be the value of $j - th$ criteria of the i-th alternative (a_i). Let assume that all parameters are to be maximized.

We use $p_j(a_i, a_k)$ denoting preferential function on criterion j as calculated below;

$$P_j(a_i, a_k) = \begin{cases} 0 & q_{ij} \leq q_{kj} \\ P(q_{ij} - q_{kj}) & q_{ij} > q_{kj} \end{cases} \qquad (6.1)$$

There are six different preference functions were adopted at Brans et al. study based on various parameters (Fig. 6.1).

Clearly, various generic parameters reflect different altitudes against preference structure and preference strength. Brans found that the linear preference function was mainly used for functional use by users led by gauss preference function. For these preference functions, the severity of the preference slowly increases from 0 to 1, while in the other preference functions, the severity of choice increases abruptly.

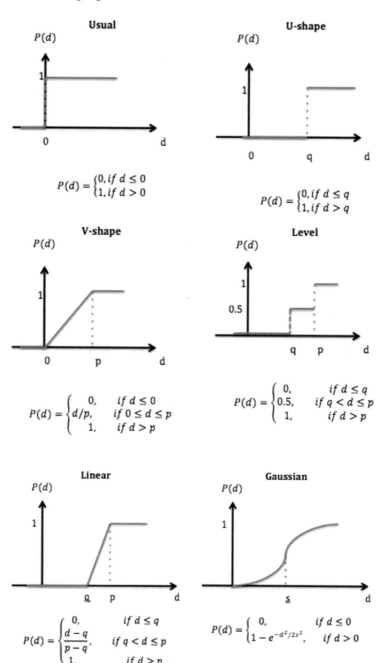

Fig. 6.1 PROMETHEE preference functions (Brans and Mareschal 2019)

The ranking results also can be obtained by decision-maker 's expectations in the PROMETHEE methods by suggesting several potential extensions for each criterion. So we can use the notion of preferred strength to incorporate various extensions to the notion to criteria. The main feature of the PROMETHEE methods is that the decision-maker will find every possible extension very clear and easily understood.

For every pair of alternatives $(a_t, a_{t'} \in A)$, we define an index of preferences for all of the parameters.

$$\pi(a_t, a_{t'}) = \sum_{k=1}^{K} w_k.[p_k(f_k(a_t) - f_k(a_{t'}))], \; AXA \rightarrow [0, 1] \quad (6.2)$$

$\pi(a_t, a_{t'})$ indicates the measure of the preference of a_t over $a_{t'}$ while w_k indicates the importance weights of the criteria k: the closer $\pi(a_t, a_{t'})$ to 1, the greater the preference.

6.2 PROMETHEE 1

The positive outranking flow $\Phi^+(a_t)$ and the negative outranking flow $\Phi^-(a_t)$ of the alternative a_t, can be obtained based on the preference indices and can be calculated by the following formulas:

$$\Phi^+(a_t) = \frac{1}{n-1} \sum_{\substack{t'=1 \\ t' \neq t}}^{n} \pi(a_t, a_{t'}) \quad (6.3)$$

$$\Phi^-(a_t) = \frac{1}{n-1} \sum_{\substack{t'=1 \\ t' \neq t}}^{n} \pi(a_{t'}, a_t) \quad (6.4)$$

The positive outranking flow $\Phi^+(a_t)$ shows how much the alternative a_t dominates the other alternatives, while the negative outranking flow $\Phi^-(a_t)$ shows how much the alternative a_t dominated by the other alternatives.

a_t is preferred to $a_{t'}$ $(a_t P a_{t'})$ when it satisfies the following condition:

$$(a_t P a_{t'}) \; if;$$
$$\begin{cases} \Phi^+(a_t) > \Phi^+(a_{t'}) \; and \; \Phi^-(a_t) \leq \Phi^-(a_{t'}) \\ \Phi^+(a_t) = \Phi^+(a_{t'}) \; and \; \Phi^-(a_t) < \Phi^-(a_{t'}) \end{cases} \quad (6.5)$$

a_t is indifferent to $a_{t'}$ $(a_t I a_{t'})$ when they both have an equal positive and negative flows:

$$(a_t I a_{t'}) \text{ if:} \Phi^+(a_t) = \Phi^+(a_{t'}) \text{ and } \Phi^-(a_t) = \Phi^-(a_{t'}) \tag{6.6}$$

a_t is incomparable to $a_{t'}$ $(a_t R a_{t'})$ if;

$$\begin{cases} \Phi^+(a_t) > \Phi^+(a_{t'}) \text{ and } \Phi^-(a_t) > \Phi^-(a_{t'}) \\ \Phi^+(a_t) < \Phi^+(a_{t'}) \text{ and } \Phi^-(a_t) < \Phi^-(a_{t'}) \end{cases} \tag{6.7}$$

This is the partial relation of the PROMETHEE I. This offers a matrix for the decision-maker on which certain actions are equal and some are not. This knowledge can be used fruitfully for decision taking in practical applications.

6.3   PROMETHEE 2

Suppose the decision-maker needs to propose the total preorder (complete listing without incomparabilities). We can then consider the net-flow $\Phi^{net}(a_t)$ for each alternative $a_t \in A$ applying to following Formula:

$$\Phi^{net}(a_t) = \Phi^+(a_t) - \Phi^+(a_t) \tag{6.8}$$

In comparison to other MCDM methods, PROMETHEE is a successfully used approach. The advantages of PROMETHEE include that it is a user-friendly method that can be perfectly applied to real-life problem structures. Both PROMETHEE I and II as whole enable the ranking of the alternatives respectively, while still providing simplicity (Ozsahin 2019).

The PROMETHEE II method arranges objects from the best (more precisely, from the most preferred) to the worst (to the least preferred). To do this, the net flows are calculated for each alternative and then ordered in descending order. The higher Φ^{net} is the better alternative (Ozsahin et al. 2019). And based on the net flow values the order of the selection problem can be obtained perfectly.

References

Brans J, De Smet Y (2016) PROMETHEE Methods. Multiple Criteria Decision Anal 187–219

Brans J, Mareschal B (2019) Promethee methods. Cin.ufpe.br 2019. [Online] Available https://www.cin.ufpe.br/~if703/aulas/promethee.pdf. Accessed 04 Oct 2019

Brans JP (1982) L'ingénierie de la décision: élaboration d'instruments d'aide à la décision. La méthode PROMETHEE, Presses de l'Université Laval

Ozsahin D, Isa N, Uzun B, Ozsahin I (2018) Effective analysis of image reconstruction algorithms in nuclear medicine using fuzzy PROMETHEE. in. Adv Sci Eng Technol Int Conf (ASET) 2019:1–5

Vincke P, Brans J-P (1985) A preference ranking organization method: the PROMETHEE method for multiple criteria decision-making. Manage Sci 31:647–656

Chapter 7
Vlse Criterion Optimization and Compromise Solution in Serbian (VIKOR)

Berna Uzun and Dilber Uzun Ozsahin

Abstract From the Serbian language, VIsekriterijumska optimizcija i KOmpromisno Resenje (VIKOR) is a way of finding a compromise ranking created by Serafim Oprikovic. VIKOR is a method that determines the superior value in comparing two alternatives for a final set of other actions that must be ranked and selected between the criteria, and resolves a discrete multicriteria problem with disparate and conflicting aspects. VIKOR pays more attention to demanding and choosing one of the best from the set of variables and determines compromise difficulties with conflicting aspects that can help decision makers to show the best alternative. A compromise conclusion is the final conclusion among the alternatives, closer to ideal solution. The VIKOR and TOPSIS methods are based on distance calculation to ideal solution of the decision matrix, but the compromise solution in VIKOR is guided by mutual concessions, while in TOPSIS the best conclusion is guided by the minimum distance from positive ideal solution (PIS) and the farthest distance from negative ideal solution (NIS). PIS is considered to be a type that consists of the best ratings between all considered criteria or attributes. On the other hand, NIS is considered a candidate that does not contain the best ratings between all the criteria considered. In this study the detailed information of the VIKOR technique will be presented.

B. Uzun (✉) · D. Uzun Ozsahin
DESAM Institute, Near East University, Nicosia, Turkish Republic of Northern Cyprus, Turkey
e-mail: berna.uzun@neu.edu.tr

D. Uzun Ozsahin
e-mail: dilber.uzunozsahin@neu.edu.tr

B. Uzun
Department of Mathematics, Near East University, Nicosia, Turkish Republic of Northern Cyprus, Turkey

D. Uzun Ozsahin
Department of Biomedical Engineering, Near East University, Nicosia, Turkish Republic of Northern Cyprus, Turkey

Medical Diagnostic Imaging Department, College of Health Science, University of Sharjah, Sharjah, United Arab Emirates

D. Uzun Ozsahin et al. (eds.), *Application of Multi-Criteria Decision Analysis in Environmental and Civil Engineering*, Professional Practice in Earth Sciences,
https://doi.org/10.1007/978-3-030-64765-0_7

Keywords Decision making · Multi criteria decision making · Compromise solution · VIKOR

7.1 Introduction

VIKOR is a multi-criteria decision making model which was first developed by Yu (1973) and widely used at the present time. This method is used for ranking and selection of a set of alternatives by considering different criteria's. In this method, the multi-criteria ranking index is introduced depending on the specific measure of closeness to the ideal solution (Zhicheng et al. 2019). VIKOR gives multi-criteria optimization and compromises solution to a complex problem. This method can be considered linear normalization.

The main part of the VIKOR is defining the positive and negative ideal points of the alternatives. VIKOR method, like any other method of MCDM analysis it follows some steps for the solution of a particular problem.

The VIKOR technique has turned into quite important multi-criteria decision making techniques because of its simplicity in calculation and accuracy in solutions. VIKOR provides the decision maker achieve the best decision and take an action by selecting and ranking options from decision matrix, and it also provides the compromise solution to the related problem where the conflicting criteria arise. It provides the compromise ranking list comprising peculiar measure starting from the closest solution to the ideal solution. However, some of its hybrid combinations, such as comprehensive VIKOR, fuzzy VIKOR, regret theory-based VIKOR, modified VIKOR and interval VIKOR techniques have also been recently established, keeping in notice the nature of multi criteria decision making analysis and the decision maker's requiement (Ali Jahan 2011).

7.2 Mathematical Framework of the VIKOR Method

The VIKOR technique has the following steps (Sayadi et al. 2009):
Step 1. Establish the decision matrix as seen in Table 7.1.

Table 7.1 General form of a decision matrix for the use of VIKOR technique

Alternative/Criteria	Criterion 1	Criterion 2	...	Criterion N
Alternative 1	f_{11}	f_{12}	...	f_{1N}
Alternative 2	f_{21}	f_{22}	...	f_{2N}
...	...	...	...	...
Alternative M	f_{M1}	f_{M2}	...	f_{MN}

Step 2. Calculate the best (f_j^*) and the worst (f_j^-) values of each criteria.

In this step the best (f_j^*) and the worst (f_j^-) values should be collected for each evaluation factor. If the aim of the criterion $-j$ is defined as maximum (for ex. if it shows the quality of a product) the best value will be calculated with the Formula (7.1):

$$f_j^* = \max_i f_{ij} \tag{7.1}$$

If the aim of the criterion-j is defined as minimum (for ex. if it shows the cost of a product) the best value will be calculated with the Formula (7.2):

$$f_j^* = \min_i f_{ij} \tag{7.2}$$

Step 3. Calculate the Utility (S_i) and Regret (R_i) measures as in Formulas (7.3) and (7.4) where w_j denotes the weights of the criteria, which represents the relative importance degrees:

$$S_i = \sum_{j=1}^n w_j \left[\frac{f_j^* - f(ij)}{f_j^* - f_j^-} \right] \tag{7.3}$$

$$R_i = \max_i \left(w_j \left[\frac{f_j^* - f(ij)}{f_j^* - f_j^-} \right] \right) \tag{7.4}$$

Step 4. Calculating the value of Q_i:

Q_i values can be calculated with the relation shown in Formula (7.5):

$$Q_i = v \left[\frac{S_i - \min(S_i)}{\max(S_i) - \min(S_i)} \right] + (1 - v) \left[\frac{R_i - \min(R_i)}{\max(R_i) - \min(R_i)} \right] \tag{7.5}$$

Here v can take any value from 0–1 and shows the weights of the strategy that provides the maximum group utility, while $(1 - v)$ shows the weight of the individual regret. The value of the v can be considered as 0.5.

Step 5. Rank the alternatives based the Q_i, R_i and S_i values in decreasing order. This will provide a three lists to the decision maker. And an alternative with the minimum Q_i provides the best solution (A') if it satisfies the following conditions:

Condition 1: (Acceptable advantage).

This condition states that there is a difference between the best and the closest to the best option.

$$Q(A'') - Q(A') \geq DQ$$

where $Q(A'')$ has the second minimum values of Q_i and $DQ = 1/(m - 1)$ where m denotes the number of the alternatives then A'.

Condition 2: (Acceptable stability).

A' must have the best value/s of the R_i and/or S_i amongst the other alternatives.

In this method, if one of the conditions is not d-satisfied, then the compromise solutions set can be proposed as below:

If only the second condition is not satisfied; A' and A''

If the first condition is not satisfied; A', A'',...A^M where M determined as the maximum decision points satisfies the condition $Q(A^M) - Q(A') < DQ$.

References

Ali Jahan MB (2011) March 3. A comprehensive VIKOR method for material selection. Retrieved from sciencedirect.net: https://www.sciencedirect.com/science/article/pii/S02613069 10006114#:~:text=In%20VIKOR%20approach%2C%20the%20compromise,agreement%20e stablished%20by%20mutual%20concessions.&text=The%20advantage%20of%20proposing% 20comprehensive,covers%20all%20objectives%20

Sayadi M, Heydari M, Shahanaghi K (2009) Extension of VIKOR method for decision making problem with interval numbers. Appl Math Model 33(5):2257–2262. https://doi.org/10.1016/j. apm.2008.06.002

Yu PL (1973) A class of solutions for group decision problems. Manage Sci 19(8):936

Zhicheng G, Liang RY, Xuan T (2019) Results in engineering VIKOR method for ranking concrete bridge repair projects with target-based criteria. Results Eng 3:100018. https://doi.org/10.1016/ j.rineng.2019.100018

Chapter 8
Fuzzy Logic and Fuzzy Based Multi Criteria Decision Analysis

Berna Uzun, Dilber Uzun Ozsahin, and Basil Duwa

Abstract Fuzzy logic is an arithmetic that allows programs on a computer to simulate the real world problems. This is a simple method to reason with uncertain, diverse and inaccurate data or knowledge. In Boolean logic, any statement is considered true or false; that is, it contains the true meaning of 1 or 0. Numerous Booleans impose strict membership requests. In fuzzy environment, every elements defined with the degree of its membership, and clear reasoning is considered as a limiting case of indicative thinking. Therefore, Boolean logic is considered a subset of fuzzy logic. People take part in the analysis of conclusions because the adoption of conclusions must take into account the subjectivity of a person, and not only apply impartial probabilistic measures. The vagueness of the situations are commonly arrises in the real world problems. In fuzzy logic it is accepted to do the necessary calculations in the environment in which goals and limits of the problem does not have the ability to be literally defined or literally presented in exact form. Zadeh proposed using the concept of fuzzy sets as a modelling tool for difficult systems that have every chance of being controlled by people, but which are difficult to literally qualify to deal with high-quality, inaccurate information or even poorly structured conclusion problems. Fuzzy based multi criteria decision analysis is the approach that allow the decision

B. Uzun (✉) · D. Uzun Ozsahin
DESAM Institute, Near East University, Nicosia, Turkish Republic of Northern Cyprus, Turkey
e-mail: berna.uzun@neu.edu.tr

D. Uzun Ozsahin
e-mail: dilber.uzunozsahin@neu.edu.tr

B. Uzun
Department of Mathematics, Near East University, Nicosia, Turkish Republic of Northern Cyprus, Turkey

D. Uzun Ozsahin · B. Duwa
Department of Biomedical Engineering, Near East University, Nicosia, Turkish Republic of Northern Cyprus, Turkey

D. Uzun Ozsahin
Medical Diagnostic Imaging Department, College of Health Science, University of Sharjah, Sharjah, United Arab Emirates

© The Author(s), under exclusive license to Springer Nature Switzerland AG 2021
D. Uzun Ozsahin et al. (eds.), *Application of Multi-Criteria Decision Analysis in Environmental and Civil Engineering*, Professional Practice in Earth Sciences,
https://doi.org/10.1007/978-3-030-64765-0_8

maker define the problem based on the fuzzy logic by fuzzification in order to give the ranking to the alternative.

Keywords Fuzzy logic · Fuzzy operations · Fuzzy based multi criteria decision making

8.1 Introduction

The Fuzzy logic comprises of a mathematical and logical variables that distinguish logical values, indicating a complete, partial and false truth, ranging from numbers 0 to 1 (Zadeh 1973). In other words, Fuzzy logic tends to analyze natural language data mathematically. The technique characterizes transitional values, which vary from real and false when evaluated.

Fuzzy logic predates to its evolution in the 1920's when it was studied and evaluated by Tarski and Lukasiewicz and later introduced and proposed by Lotfi Zadeh in 1965. Fuzzy method was introduced as a result of uncertainty in evaluation and analysis (Zadeh 1968). This process is able to give a precise and elaborate analysis. The fuzzy logic, gives a proper assessment of any introduced variable and take values from 0 to 1 range which opposes the binary logic (Uzun and Kiral 2017). It is structured into three stages: fuzzification, defuzzification and rules inferences. The fuzzy logic control (FLC) structure introduced by Zadeh, proposes procedure that assists in calculating and evaluating variables that are referred to "rules induction".

Strengths Of Fuzzy Logic (Ibarra and Webb 2016):

- The FLC can be useful for two main purposes, the commercial and practical.
- The FLC controls problems affiliated to input variations and function
- The system is not limited to a particular control outputs or inputs.
- FLC manages non-linear systems which cannot or won't be able to be managed in a mathematically approach.
- Fuzzy logic is designed to control and override other target control processes.
- FLC assists in determining engineering uncertainty

Weaknesses Of Fuzzy Logic (Ibarra and Webb 2016):

- The FLC has no ability of machine learning recognition.
- It is observed, there's a huge contradiction in recognizing fuzzy logic from probability theory.
- Fuzzy logic has no accuracy
- The system lacks the potential technique used for comparison
- An accurate testing and hardware is needed for the verification of the device or model.

8.2 Fuzzy Logic Sets

Fuzzy logic sets is the extension of the classical set. These sets are objects collected with a desired or shared characters. In other words, Classical sets and Fuzzy sets are collected in two main processes with different components ranging from 0 or 1 and 0,1 sequently (Kiral and Uzun 2017, 2019). These processes express the classical and Fuzzy sets.

8.2.1 The Mathematical Expression of Fuzzy Sets

Fuzziness is defined mathematically as a derived mathematics of logic and set theories that enable evaluation as derived and proposed by Zadeh Lofti in 1965 (Zadeh 1965). The mathematical expression is detailed as follows:

A fuzzy set $\tilde{A} \in IR$ is a set of pairs as:

$$A = \{(x, \mu_{\tilde{A}}(x)) | x \in IR\} \tag{1}$$

Therefore, $\mu_{\tilde{A}}: IR \to [0, 1]$ and $\mu_{\tilde{A}}(x)$ denotes the membership function of the A (Kiral 2018).

The fuzzy set is represented by the mathematical expression in various ways; these processes are illustrated below.

Where "U" is discrete and finite:

$$\tilde{A} = \left\{ \frac{\mu_{\tilde{A}}(x_1)}{x_1} + \frac{\mu_{\tilde{A}}(x_2)}{x_2} + \frac{\mu_{\tilde{A}}(x_3)}{x_3} + \cdots \right\} = \sum_{i=1}^{n} \frac{\mu_{\tilde{A}}(x_i)}{x_i} \tag{2}$$

where "U" is continuous and infinite:

$$\tilde{A} = \left\{ \int \frac{\mu_{\tilde{A}}(x)}{x} \right\} \tag{3}$$

The expression above show, each element of set is defined by the symbol summation, where U represented the universe of information.

8.2.2 Logical Operations of the Fuzzy Sets

The Fuzzy set operation is connected by union, complement operation and intersection on fuzzy sets. These outlined fuzzy set complements are characterized by different mathematical expressions. The following is a relationship expressed to describe the fuzzy sets, union and intersection.

Union:

$$\mu_{\tilde{A} \cup \tilde{B}}(x) = \mu_{\tilde{A}} \vee \mu_{\tilde{B}}, \quad \forall x \in U \tag{4}$$

$\vee$ represents the 'max' operation.

Intersection:

$$\mu_{\tilde{A}\cap\tilde{B}}(x) = \mu_{\tilde{A}} \wedge \mu_{\tilde{B}}, \quad \forall x \in U \tag{5}$$

$\wedge$ represents the 'min' operation.

Complement:

$$\mu'_{(\tilde{A})}(x) = 1 - \mu_{\tilde{A}}(x) \tag{6}$$

There may be some cases there;

$$\tilde{A} \cap \tilde{A}' \neq 0$$

8.2.3 Fuzzy Sets Features

Fuzzy sets are described as sets that contain variables with similar values or membership. The fuzzy sets are characterized by different properties, which are as follows:

(a) Commutativity:

The Commutativity property involves two variables and relating them together by analyzing them. Many systems contain the commutativity property; a clear example is the case of two or more inputs of a similar character. This involves fuzzy set $\tilde{A}$ and $\tilde{B}$, and states that:

$$\tilde{A} \cup \tilde{B} = \tilde{B} \cup \tilde{A} \tag{7}$$

$$\tilde{A} \cap \tilde{B} = \tilde{B} \cap \tilde{A} \tag{8}$$

(b) Associativity

This involves a particular mathematical property of a derived binary operation which cannot have effect on the given result. This involves fuzzy sets $\tilde{A}$, $\tilde{B}$ and $\tilde{C}$, and states that:

$$\tilde{A} \cup \left(\tilde{B} \cup \tilde{C}\right) = \left(\tilde{A} \cup \tilde{B}\right) \cup \tilde{C} \tag{9}$$

$$\tilde{A} \cap \left(\tilde{B} \cap \tilde{C}\right) = \left(\tilde{A} \cap \tilde{B}\right) \cap \tilde{C} \tag{10}$$

(c) Distributive Property:

The distributive property uses three fuzzy sets properties, this involves fuzzy sets $\tilde{A}$, $\tilde{B}$ and $\tilde{C}$, and states that:

$$\tilde{A} \cup \left(\tilde{B} \cap \tilde{C}\right) = \left(\tilde{A} \cup \tilde{B}\right) \cap \left(\tilde{A} \cup \tilde{C}\right) \tag{11}$$

$$\tilde{A} \cap \left(\tilde{B} \cup \tilde{C}\right) = \left(\tilde{A} \cap \tilde{B}\right) \cup \left(\tilde{A} \cap \tilde{C}\right) \tag{12}$$

(d) Idempotency:

The Idempotency technique can be expressed given fuzzy set $\tilde{A}$, it can be stated that:

$$\tilde{A} \cup \tilde{A} = \tilde{A} \tag{13}$$

$$\tilde{A} = \tilde{A} \cap \tilde{A} \tag{14}$$

(e) Identity Property:

The property illustrated mathematically using a given fuzzy set $\tilde{A}$ and a universal set U, it can be stated that:

$$\tilde{A} = \tilde{A} \cap U \tag{15}$$

$$U = \tilde{A} \cup U \tag{16}$$

And also:

$$\tilde{A} = \tilde{A} \cup \emptyset \tag{17}$$

$$\emptyset = \tilde{A} \cap \emptyset \tag{18}$$

(f) Transitivity:

This can be shown by using fuzzy sets $\tilde{A}$, $\tilde{B}$ and $\tilde{C}$, the transitivity feature states:

$$\text{If } \tilde{A} \subseteq \tilde{B} \text{ and } \tilde{B} \subseteq \tilde{C} \text{ then } \tilde{A} \subseteq \tilde{C} \tag{19}$$

(g) Involution:

The involution property is expressed using the following expression, provided a fuzzy set $\tilde{A}$:

$$\overset{\cong}{\tilde{A}} = \tilde{A} \tag{20}$$

De Morgan's Law:

The De Morgan's laws was coined from Augustus De Morgan, a mathematician. The law can be expressed as "the complement of two sets in Union having the same intersection with the complements of the two sets in intersection. This law plays a significant role in demonstrating redundancies and logical inconsistency. It states:

$$\overline{\tilde{A} \cup \tilde{B}} = \overline{\tilde{A}} \cap \overline{\tilde{B}} \tag{21}$$

$$\overline{\tilde{A} \cap \tilde{B}} = \overline{\tilde{A}} \cup \overline{\tilde{B}} \tag{22}$$

8.2.4　Membership Function

In the Fuzzy set, the membership function represents the weight of truth in the process of valuation. In other words, the membership function is defined as the expression set X and a real unit interval represented as 0,1. The membership degree of $x \in \tilde{A}$ can be denoted by $\mu_{\tilde{A}}(x)$. If $\mu_{\tilde{A}}(x) = 0$, x is not a member of the fuzzy set (Fig. 8.1).

Lofti A. Zadeh perfectly expresses the membership features of the fuzzy sets in a research, demonstrating the main features of the fuzzy sets. 'The main properties of membership description include:

- They form a distinct comparison in fuzziness
- The membership function use occurrences to solve real-life problems.

Fig. 8.1 A triangular fuzzy set (Ozsahin et al. 2020)

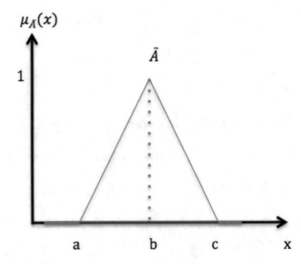

Fig. 8.2 Membership functions features (Ozsahin et al. 2020)

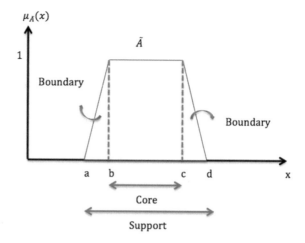

(a) Core:

The core elements of a fuzzy set $\tilde{A} \in IR$ can be determined by the formula below:

$$\mu_{\tilde{A}}(x) = 1 \qquad (23)$$

(b) Support:

The support elements of the fuzzy set $\tilde{A} \in IR$ can be determined by the formula below:

$$\mu_{\tilde{A}}(x) > 0 \qquad (24)$$

(c) Boundary:

The boundary of the fuzzy set $\tilde{A} \in IR$ can be determined by the formula below:

$$1 > \mu_{\tilde{A}}(x) > 0 \qquad (25)$$

The general features of a fuzzy set can be seen in Fig. 8.2.

8.3   Fuzzification

Fuzzification is referred to as a process by which crisp input value is converted to a fuzzy value that works through information from the knowledge base. In other words, it is seen as a process of decomposing any input or output system into one or more fuzzy sets. Fuzzification can also be a method described by changing a fuzzy set into methods such as the fuzzier set or a fuzzy set of crisp sets, respectively. It is recorded, during the process of fuzzification, the numbers of the inputs or outputs are changed into fuzzy sets. These fuzzification techniques methods are of two different ways as defined below:

(a) Support Fuzzification (s-fuzzification) Method:

These fuzzification method is obtainable through the function below:

$$\tilde{A} = \mu_1 Q(x_1) + \mu_2 Q(x_2) + \ldots + \mu_n Q(x_n) \tag{26}$$

where $Q(x_i)$ denotes the Kernel of fuzzification. This method is also applicable while keeping μ_i as constant and x_i as the values converted to a fuzzy set $Q(x_i)$.

(b) Grade Fuzzification (g-fuzzification) Method:

This procedure is nearly comparative to the s-fuzzification strategy with the distinction within the two parameters. In this procedure, x_i indicates a constant and μ_i indicates a fuzzy set.

8.4 De-fuzzification

Defuzzification is defined as the method of getting a number obtained from an output from a fuzzy set. It is utilized to exchange fuzzy interference output. In other words, defuzzification can be obtained by an algorithm of decision-making that chooses the finest esteem based on fuzzy set. It is additionally a handle that changes the fuzzy method into new method. This handle is exceptionally critical for getting a meaningful output, particularly within the designing applications. De-fuzzification can be spoken to as "adjusting it off". The following operations are accessible for the de-fuzzification:

(a) Max-Membership Method: This procedure is limited to maximum output functions. It is additionally called the height method. It is numerically characterized as the following equation.

$$\mu_{\tilde{A}}(x^*) > \mu_{\tilde{A}}(x), \quad \forall x \epsilon X \tag{27}$$

where x^* is the de-fuzzified output.

(b) Centroid Method: This strategy is additionally recognized as center of area method. The output can be obtained with the equation below:

$$x^* = \frac{\int \mu_{\tilde{A}}(x) \cdot x \mathrm{d}x}{\int \mu_{\tilde{A}}(x) \cdot \mathrm{d}x} \tag{28}$$

(c) Weighted Average Method: In this procedure, the membership function is weighted with its maximum membership degree. The following equation expresses the weighted average method technique, respectively.

$$x^* = \frac{\int \mu_{\tilde{A}}(x) \cdot x \mathrm{d}x}{\int \mu_{\tilde{A}}(x) \cdot \mathrm{d}x} \tag{29}$$

(d) Mean-Max Membership: This method can be determined as the middle of the maxima. The output x^* is expressed by the formula below:

$$x^* = \frac{\sum_{i=1}^{n} \bar{x}_i}{n} \tag{30}$$

8.5 Fuzzy Sets Algebraic Operations

The main fuzzy algebraic operations used can be summarized as follows

(a) Algebraic Product:

$$\tilde{A} \cdot \tilde{B} \Leftrightarrow \mu_{\tilde{A} \cdot \tilde{B}} = \mu_{\tilde{A}} \cdot \mu_{\tilde{B}} \tag{32}$$

(b) Algebraic Sum:

$$\tilde{A} + \tilde{B} \Leftrightarrow \mu_{\tilde{A} + \tilde{B}} = \mu_{\tilde{A}} + \mu_{\tilde{B}} - \mu_{\tilde{A}} \cdot \mu_{\tilde{B}} \tag{33}$$

(c) Bounded-Sum:

$$\tilde{A} \oplus \tilde{B} \Leftrightarrow \mu_{\tilde{A} \oplus \tilde{B}} = 1 \wedge (\mu_{\tilde{A}} + \mu_{\tilde{B}}) \tag{34}$$

(d) Bounded-Difference:

$$\tilde{A} \ominus \tilde{B} \Leftrightarrow \mu_{\tilde{A} \ominus \tilde{B}} = 0 \vee (\mu_{\tilde{A}} - \mu_{\tilde{B}}) \tag{35}$$

(e) Bounded- Product:

$$\tilde{A} \odot \tilde{B} \Leftrightarrow \mu_{\tilde{A} \odot \tilde{B}} = 0 \vee (\mu_{\tilde{A}} + \mu_{\tilde{B}} - 1) \tag{36}$$

where the following symbols; $\vee, \wedge, +, -$ denote the max, min, arithmetic sum and arithmetic difference, respectively.

However, the extension of the multi criteria decision analysis method in a fuzzy environment can be achieved by expressing the importance weights of the criteria and ratings for the linguistic variables or fuzzy variables. A linguistic variable is a variable whose meanings are considered linguistic definitions. The concept of a linguistic variable can be quite useful in cases that are very complex or very badly assigned so that they can reasonably be described in classical quantitative expressions (Balioti et al. 2018). However by applying one of the defuzzification technique, the decision maker could apply to one of the multi criteria decision making techniques to solve the problem under the fuzzy condition.

References

Balioti V, Tzimopoulos C, Evangelides C (2018) Multi-criteria decision making using topsis method under fuzzy environment. Appl Spillway Select Proc 2(11):637. https://doi.org/10.3390/procee dings2110637

Ibarra L, Webb J (2016) Advantages of fuzzy control while dealing with complex/ unknown model dynamics: a quadcopter example

Kiral E (2018) Modeling brent oil price with markov chain process of the fuzzy states. Pressacademia 5(1):79–83. https://doi.org/10.17261/pressacademia.2018.785

Kiral E, Uzun B (2017) Forecasting closing returns of borsa istanbul index with markov chain process of fuzzy states. Pressacademia 4(1):15–24. https://doi.org/10.17261/pressacademia.201 7.362

Ozsahin DU, Uzun B, Ozsahin I, Mubarak MT, Musa MS (2020) Fuzzy logic in medicine biomedical signal processing and artificial intelligence in healthcare Academic Press. In: Zgallai W (ed) Series developments in biomedical engineering and bioelectronics 153–182

Uzun B, Kıral E (2017) Application of markov chains-fuzzy states to gold price. Proc Comput Sci 120:365–371. Available https://doi.org/10.1016/j.procs.2017.11.251

Uzun B, Kıral E (2019) Evaluating US dollar index movements using markov chains-fuzzy states approach. In: Aliev R, Kacprzyk J, Pedrycz W, Jamshidi M, Sadikoglu F (eds) 13th International conference on theory and application of fuzzy systems and soft computing—ICAFS-2018, ICAFS 2018 Advances in intelligent systems and computing, vol 896, Springer, Cham

Zadeh L (1965) Fuzzy sets. Inf Control 8(3):338–353. Available: https://doi.org/10.1016/s0019-9958(65)90241-x

Zadeh L (1973) Outline of a new approach to the analysis of complex systems and decision processes. IEEE Trans Syst Man Cybernet 3(1):28–44. Available https://doi.org/10.1109/tsmc.1973.540 8575

Zadeh L (1968) Fuzzy algorithms. Inf Control 12(2):94–102. Available https://doi.org/10.1016/s0019-9958(68)90211-8

Chapter 9
Predict Future Climate Change Using Artificial Neural Networks

Hamit Altıparmak, Ramiz Salama, Hüseyin Gökçekuş, and Dilber Uzun Ozsahin

Abstract In Artificial Neural Networks (ANN) with feedback, the output of at least one cell is given as input to itself or to other cells, and feedback is usually done via a delay element. Feed-back can be between cells in a layer or between cells between layers. With this structure, the feedback ANN shows dynamic nonlinear behavior. Therefore, feedback ANN structures can be obtained in different structures and behaviors depending on the type of feedback. There have been many studies documenting the increase in the average global temperature in the last century. The consequences of a continuous rise in global temperature will be significant. The rising sea levels and increasing frequency of extreme weather events will affect billions of people. Neural Net-work Performance: We used a data table comprising 8 rows and 303 columns as input. We used a feedback neural network consisting of 1 hidden layer and 10 neurons. Results: Training 90.172%, Validation 84.859%, Test 81.697%, All 87.945%. The effects of climate change have already been observed and will become more apparent in the future. With the contribution of all countries,

H. Altıparmak · R. Salama (✉)
Department of Computer Engineering, Near East University, Nicosia, Turkish Republic of Northern Cyprus, Turkey
e-mail: ramiz.salama@neu.edu.tr

H. Altıparmak
e-mail: hamit.altiparmak@neu.edu.tr

D. Uzun Ozsahin
DESAM Institute, Near East University, Nicosia, Turkish Republic of Northern Cyprus, Turkey
e-mail: dilber.uzunozsahin@neu.edu.tr

H. Gökçekuş
Faculty of Civil and Environmental Engineering, Near East University, Nicosia, Turkish Republic of Northern Cyprus, Turkey

D. Uzun Ozsahin
Department of Biomedical Engineering, Near East University, Nicosia, Turkish Republic of Northern Cyprus, Turkey

Medical Diagnostic Imaging Department, College of Health Science, University of Sharjah, Sharjah, United Arab Emirates

© The Author(s), under exclusive license to Springer Nature Switzerland AG 2021
D. Uzun Ozsahin et al. (eds.), *Application of Multi-Criteria Decision Analysis in Environmental and Civil Engineering*, Professional Practice in Earth Sciences,
https://doi.org/10.1007/978-3-030-64765-0_9

the negative impacts of climate change need to be identified. In this way, strategies to combat potential problems caused by future climate change can be established

Keywords Environmental problems · Climate change · Global warming · Artificial neural network · Prediction

9.1 Introduction

The prediction of future climate change is the most important attribute to forecast because most industries as well as agricultural sectors are largely dependent on climate conditions. Since the ancient times, climate prediction has been one of the fascinating and interesting domains. It is used to predict and warn about various natural disasters that are caused due to changes in climate conditions. Due to the confusing nature of the atmosphere, greater computational power is required for solving complex equations related to the prediction of changes in atmospheric conditions. Climate forecasting may be less accurate because of the difference between historical data and future events. With the help of these models we can minimize this error to predict the most correct outcome.

The steps involved in predicting the climate are as follows:

1. Data collection, such as maximum and minimum temperature
2. Data assimilation.
3. Data analysis
4. Numerical climate prediction

The effects of climate change include high average temperatures, extreme and frequent climate events and rising sea levels, which are expected to lead to an increase in disease and mortality, as well as negative impacts on safe food supply, clean water and sanitation.

Rising sea levels are threatening access to land in coastal areas, particularly low-lying islands. Land used for agriculture will no longer be usable, as saltwater contaminates soil and fresh water supplies. People are forced to migrate inland, causing health issues including increased infectious diseases.

Rising temperatures and other effects of climate change are creating fertile ground for disease-carrying insects. Mosquitoes, which spread diseases including malaria, dengue and Zika, are particularly sensitive to changes in temperature and humidity.

9.2 Aim of the Study

The aim of this study is to create an artificial neural network that predicts future climate changes and their effects by using climate change data from previous years. With the development of the artificial neural network, possible future situations will be predicted accurately and efficiently.

9.3   Related Work

An investigation was carried out for the problem of ensemble learning in raster classification. This problem is important in many applications, such as classification in medical image processing and land cover classification in remote sensing. The problem is challenging due to the effect of class ambiguity from spatial heterogeneity.

The artificial neural network approach can be used to solve nonlinear problems of regression that arise in environmental modeling, which includes forecasting for a short term period in addition to rainfall-runoff modeling and atmospheric concentrations that leads to pollution. The aim was to review the existing methodology for estimating predictive uncertainty as environmental datasets are redundant and noisy, which describes how a predictive distribution model can be used to assess the impacts occurred due to the climate change and to improve the different decision to avoid the different impacts of it (Cawley et al. 2007).

An investigation was carried out to determine the relationship between different metrological factors and Newcastle disease and the different key factors that cause this disease were determined. To achieve this, a Newcastle disease forecasting model was built and a Back Propagation neural network classification technique was applied for animal disease forecasting in their research. For discovery covariate which is based on a Poisson description with high-frequency Bayesian framework and Sparse regression model of hierarchical Bayesian was used to identify the covariates that affect the precipitation frequency which are collected from the various observations at different stations over many different climatologically regions which are in the U.S. continental (Debasish Das et al. 2014).

9.4   Artificial Neural Network

Artificial neural networks first learn by collecting the obtained data in the system, then obtain results using the information they have learned against the samples that are never presented to the network. Because of these learning and generalization features, artificial neural networks have found a wide range of application opportunities in many fields of science and have demonstrated the ability to solve complex problems successfully. According to another definition, artificial neural networks are parallel and distributed information processing structures that are inspired by the human brain and are connected to each other by means of predominant connections where each of which has its own memory; in other words, they are computer programs that mimic biological neural networks.

9.4.1 Training Sets and Training Process

The training process is not random and must be planned in advance. The input data should not be randomly transmitted to the artificial neural network. Preparing the data to be trained in the network in advance and entering the network in sequential order provides better results. The regular data set for training the network is called the training set, which should contain as much data that the network may encounter later as possible. After the network has been trained, acceptable answers can be received from the network when a series of data is entered in the network that is not included in the training set. This capability of the network increases in direct proportion to the diversity of data in the training set.

9.4.2 Scaling of Input and Output Data

In a neural network that uses a sigmoid function, the network can be given numbers between 0 and 1 or only between these two values. However, the network may be required to use large number to achieve this. The input and output data between any 2 values are multiplied by two coefficients. Thus, as the network continues to operate with values between 0 and 1, the input-outputs can be exchanged at the desired intervals.

9.4.3 Minimum–Maximum Normalization (Min–Max Normalization)

In this method, the largest and smallest values in a group of data are handled. All other data is normalized to these values. The aim is to normalize the smallest value to 0 and the largest value to 1, and to spread all other data over this 0–1 range.

Various studies have documented the increase in the average global temperature in the last century. The consequences of a continuous rise in global temperature will be significant. Rising sea levels and increased frequency of extreme weather events will affect billions of people.

The dataset contains data from the past 25 years.

9.5 Neural Network Performance

We used a data table comprising 8 rows and 303 columns as input. We used a feedback neural network consisting of 1 hidden layer and 10 neurons (Fig. 9.1).

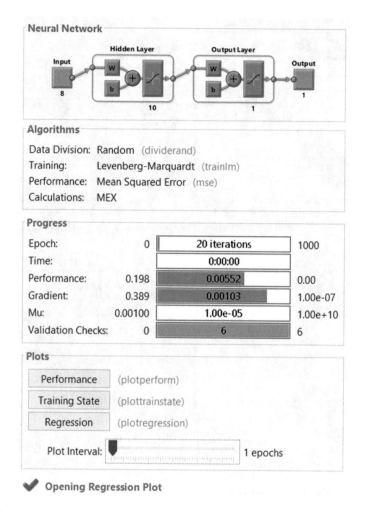

Fig. 9.1   Neural network performance

9.6   Performance Result and Conclusion

The effects of climate change have already been observed and will become more apparent in the future. With the contribution of all countries, the negative impacts of climate change need to be identified. In this way, strategies to combat potential problems caused by climate change in the future can be established.

Recently, there has been a rapid increase in systems predicting future situations. Artificial neural networks can be safely applied to solve such problems. The advantages are that they can solve problems under incomplete information conditions without the analytical relationship knowledge between input and output data.

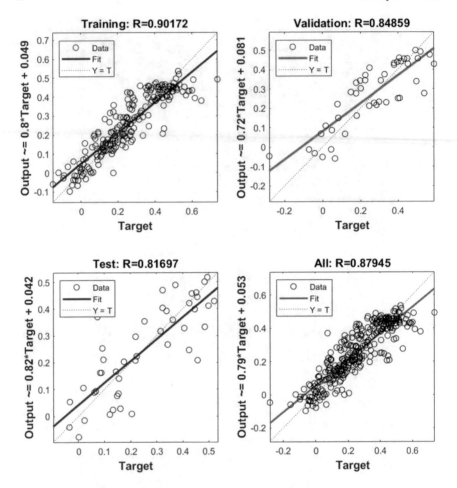

Fig. 9.2 Neural network performance

With this application, we have seen that we can predict future climate change situations with neural networks that produce more successful results. As can be seen from the database we used, all the factors that make up the atmosphere can have an impact on climate change. As a result, we used a dataset consisting of MEI, CO_2, CH_4, N_2O, CFC-11, CFC-12, TSI and Aerosols. It is observed that our training success is 90.172%, validation success is 84.859%, test success is 81.697% and overall success is 87.945% (Fig. 9.2; Table 9.1).

Table 9.1 Performance result

Training	90.172%
Validation	84.859%
Test	81.697%
All	87.945%

References

Abellatif M, Vatner SF, Vatner DE, Sadoshima J (2015) U.S. Patent No. 9,051,568. Washington, DC, U.S. Patent and Trademark Office

Baboo DSS, Shereef IK (2011) An efficient temperature prediction system using BPN neural network. Int J Environ Sci Develop 2(1):2010–2264

Cawley GC, Talbot NL, Girolami M (2007) Sparse multinomial logistic regression via bayesian l1 regularisation. In: Advances in neural information processing systems, pp 209–216

Chauhan S, Shrivastava RK (2009) Performance evaluation of reference evapotranspiration estimation using climate based methods and artificial neural networks. Water Resour Manage 23(5):825

Choudhary PR, Thakur H, Murade A, Venkatesan M (2016) Climate change prediction using artificial neural network. Int J Appl Eng Res 11(3):1954–1959

Das D, Bakshi S, Bhattacharya P (2014) Dyeing of sericin-modified cotton with reactive dyes. J Textile Inst 105(3):314–320

Ferrari S, Ruggieri P, Cefalo G, Tamburini A, Capanna R, Fagioli F, Alberghini M (2012) Neoadjuvant chemotherapy with methotrexate, cisplatin, and doxorubicin with or without ifosfamide in nonmetastatic osteosarcoma of the extremity: an Italian sarcoma group trial ISG/OS-1. J Clin Oncol 30(17):2112–2118

Ghorbani MA, Khatibi R, Hosseini B, Bilgili M (2013) Relative importance of parameters affecting wind speed prediction using artificial neural networks. Theoret Appl Climatol 114(1–2):107–114

Joseph J, Ratheesh TK (2013) Rainfall prediction using data mining techniques. Int J Comput Appl 83(8)

Karpatne A, Khandelwal A, Boriah S, Kumar V (2014) Predictive learning in the presence of heterogeneity and limited training data. In: Proceedings of the 2014 SIAM international conference on data mining, Society for industrial and applied mathematics, pp 253–261

Mehra P, Wah BW (1992) Adaptive load-balancing strategies for distributed systems. In: Proceedings of the second international conference on systems integration, IEEE, pp 666–675

Nikam VB, Meshram BB (2013) Modeling rainfall prediction using data mining method: a Bayesian approach. In: 2013 Fifth international conference on computational intelligence, modelling and simulation, IEEE, pp 132–136

Raje D, Mujumdar PP (2010) Reservoir performance under uncertainty in hydrologic impacts of climate change. Adv Water Resour 33(3):312–326

Sencan H, Chen Z, Hendrix W, Pansombut T, Semazzi F, Choudhary A, Samatova NF (2011) Classification of emerging extreme event tracks in multivariate Spatio-temporal physical systems using dynamic network structures: application to hurricane track prediction. In: Twenty-second international joint conference on artificial intelligence

Chapter 10
Superior Types of Bamboo as a Construction Material with MCDM Methods

Aizhan Syidanova, Huseyin Gokcekus, and Dilber Uzun Ozsahin

Abstract Bamboo is one of the oldest types of material that can be used in many areas of human life. In the construction industry, bamboo is quite common, but not every species can be used. In construction industry important part is how to use bamboo and which treatment to use, how not to harm environment with methods of treatment. The considered types of bamboo can not only be used in construction, but also in various other areas of construction and design. Tables showing the types of bamboo and their comparison, the types of construction approaches frequently used by architects and the important criteria for bamboo as a building material are presented. MCDM methods was described in work and was selected as the multi-criteria method as it identifies a solution that takes into account several criteria based on a comparison of each aspect regarding the selected criterion. Bamboo has significant potential as a building material and there is confidence that it is an environmentally friendly material. Confidence in choosing the right type of bamboo is just as important as constructing using the bamboo.

A. Syidanova (✉)
Department of Architecture, Near East University, Nicosia, Turkish Republic of Northern Cyprus, Turkey
e-mail: aizhansyidanova@gmail.com

H. Gokcekus
Faculty of Civil and Environmental Engineering, Near East University, Nicosia, Turkish Republic of Northern Cyprus, Turkey
e-mail: huseyin.gokcekus@neu.edu.tr

D. Uzun Ozsahin
DESAM Institute, Near East University, Nicosia, Turkish Republic of Northern Cyprus, Turkey

Department of Biomedical Engineering, Near East University, Nicosia, Turkish Republic of Northern Cyprus, Turkey

Medical Diagnostic Imaging Department, College of Health Science, University of Sharjah, Sharjah, United Arab Emirates

D. Uzun Ozsahin
e-mail: dilber.uzunozsahin@neu.edu.tr

© The Author(s), under exclusive license to Springer Nature Switzerland AG 2021
D. Uzun Ozsahin et al. (eds.), *Application of Multi-Criteria Decision Analysis in Environmental and Civil Engineering*, Professional Practice in Earth Sciences, https://doi.org/10.1007/978-3-030-64765-0_10

65

Keywords Types of bamboo · Bamboo material · Construction material ·
Treatment of bamboo · Using of bamboo · Sustainable architecture · Multi-criteria
decision-making (MCDM)

10.1 Introduction

People can understand that mass cutting of trees will not lead the future of construction to good ecology, scientists looking for answers to the question "What materials can be used without harming the environment?".

Given the global environmental situation on the planet, scientists are constantly looking for new ways to conserve greenery on the continents, to introduce new, and safer and more progressive production technologies. As an old material which ancients used before bamboo is versatile grass which people can use in different industries with applying new creative way of using it. In the twenty-first century, bamboo as an innovation in production takes a big step forward, and makes a big influence in people's daily lives.

Bamboo fibers favorably differ from many others, because in addition to all the advantages of natural fibers have unique properties as hypoallergenicity; antibacterial properties that persist even after prolonged use; high wear resistance, resistance to burnout; hygroscopicity; the ability to retain ultraviolet rays; natural antistatic effect; has potential of antioxidant.

Bamboo is a type of grass that grows in tropical and subtropical climates, in natural conditions in Asia, Europe, America, Africa, and Australia as well in Oceania. It is considered a twenty-first century material it is durable, light and does not absorb water. Bamboo is a utilitarian plant, depending on the treatment, can vary in properties. It can be firm and stable, also soft and elastic. Some use it in building houses, covering floors, and others for sewing textiles. In the twenty-first century, this material used in woodworking industry; pulp and paper industry; textile industry; bioenergetics from grass; food and beverages; automotive industry; high technology area; agriculture; also being used as packaging and as an ingredient in cosmetic products.

In everyday life thanks to the ecological properties of raw bamboo can be found in your kitchen in the form of wooden spoons, baskets and bowls, which are usually varnished and painted. Big competition bamboo makes up plastic, as it is environmentally friendly. Covers, keyboards and even computer mice are made of polished wood. Bamboo as wood is used in furniture, huts, sailing ships, masts, carpets, curtains, pots, paper, clothes, and many more applications (Jamatia 2015). Bamboo coal creates clean air and reduces residues of contaminants; bicycles from bamboo are still produced and new types are being created.

What kinds of bamboo can be used as building materials? How and where can bamboo be used and in which area?

Theoretical analysis of the material of bamboo is based on a deep comprehension of the fact that in architecture, there is no material that does not cause harm to the environment. As a plant, bamboo is a fast-growing and strong material; however, it requires special care before it can be used in the construction sector. The bamboo plant has 1,300 types, but not all can be used as a building material. A table showing the types of bamboo that can be used in construction has been compiled for more in-depth study.

10.1.1   Bamboo as a Plant

Depending on the genus, the growth of bamboo varies from 20 to 50 cm to 40 m. It is one of the fastest-growing plants in the world and can grow up to 0.75 m per day. Curly bamboo lianas reach up to 120 cm in length. The development of bamboo can depend on climatic conditions, where the growth of tropical species of bamboo may depend on the level of humidity, during the rainy season, the activity of growth increases. Other groups of bamboo species are designed for subtropical and temperate climate with cold or cool winters, and their growth activity is controlled by seasonal conditions. In addition, if herbaceous bamboo grows exclusively in the tropics, then some richer species of bamboo meet and feel good in the more northern cold areas. In the eastern Andes, one can find bamboo Chusquea Aristata at an altitude of 4700 m above sea level, which here form impenetrable thickets that meet up to the snowboarder. In the Himalayas, at an altitude of 3800 m. Currently, Bambusa Metake from Japan and several species of bamboo from China are perfectly acclimatized and grow in Central Europe. Bamboo are monocarpic plants meaning that after flowering, they perish. Bamboo flowering occurs very rarely every 60–130 years, after the whole forest blooms, it comes to the doom of a bamboo grove (Janssen 2000).

10.1.2   Types of Bamboo and Their Feature Specifications

Bamboo (Bambusa) belongs to the same family as Cereals, and there are about 115 genera of bamboo and 1300 species. Almost all varieties of these plants come from the Far East. Bamboo is an evergreen plant, that has a stiffened stalk, on which lanceolate leaves and solitary flowers, or paniculate inflorescence, are located at the top. Bamboo are monocarpic plants meaning that after flowering, they perish. The bamboo trunk is a long natural "column", comprised of a uniformly thick and dense cylinder, with a smooth, siliceous, naturally polished surface (Panday and Suneel 2007). At regular intervals, it is interrupted by annular nodes, each interstitial site being a completely isolated waterproof vessel. Its size and shape are highly diverse in terms of the proportions and length of individual internodes, the thickness and strength of partitions, in straightness, smoothness, and durability. Furthermore, some bamboo heavy and other are light, and they have different wall thicknesses. Although it is very strong, bamboo is at the same time elastic; in both and dry forms, it has no specific smell or taste (Patel 2015).

10.1.3 The Initial Use of Bamboo

More than 5000 years ago, people used bamboo for projects including shooting, construction, weaving, books and paper. The first known study of bamboo was record, during the Jin Dynasty, China, in a book, which included 61 different types; there was biological description and cultivation method. The reason why bamboo has such an ancient history in China and widespread use, because over 400 species of grass grow on the territory of this country, ranking first in the number of growing bamboo throughout the world.

In 206 BC, the main craft of the Khan Dynasty was the production of paper, and they produced it precisely from bamboo. It was profitable from 3000 kg of bamboo to make 1000 kg of paper. There was one drawback with this paper; it was heavy due to mixing with other materials (amaZulu Inc 2014).

1037–1101 AD Bamboo was used for all kinds of items necessary for life. This plant was used for firewood, paper, rafts, tiles, hats, shoes, capes and much more (amaZulu Inc 2014).

In 1486 during the Ming dynasty in China bamboo made coal, which is a sustainable and healthy alternative to traditional coal. Bamboo charcoal creates clean air and reduces residues of contaminants (amaZulu Inc 2014).

Patented the first bamboo bicycles in 1894 in England. Bicycles from bamboo are still produced and new types are being created (amaZulu Inc 2014).

During the Second World War, Gucci designers showed creativity and created the Gucci Bamboo Bag. Since the resources were not enough, the company used bamboo as handles for bags. Developers of Gucci have developed a patented method of heating and bending bamboo, this was done in order for the handle of the bag to retain its shape after cooling and fastening (amaZulu Inc 2014).

1997 The international bamboo and rattan network exists for the development of environmental sustainability using bamboo and rattan. It was founded in 1997 and believes that stability can be achieved in poor areas around the world if one finds ways to use these plants creatively (amaZulu Inc 2014).

Medicine is another area where one cannot do without the beneficial properties of bamboo. It helps with heartburn as it has the property of cooling, with fainting, treatment of epilepsy, reduces phlegm in the throat. In India and Tibet apply during fever and spasms, and is also part of aphrodisiacs and tonics.

In everyday life thanks to the ecological properties of raw bamboo can be found in your kitchen in the form of wooden spoons, baskets and bowls, which are usually varnished and painted. Big competition bamboo makes up plastic, as it is environmentally friendly. Covers, keyboards and even computer mice are made of polished wood.

10.1.4 *Sustainability of Bamboo*

Sustainability of in construction area (energy, water, raw materials), allocate waste (construction and demolition) and allocate potentially harmful emissions into the atmosphere. Architects, designers and builders have to face the difficult task of meeting the needs for new buildings and structures that are affordable, safe, healthy and productive while minimizing any negative consequences for society, the environment and the economy. Under the best circumstances, building structures should lead to positive results in all three areas.

Sustainable architecture is the construction of buildings that create the best living conditions for a person. Improving and minimizing the use of harmful resources in architecture and design. This effect on building materials and construction methods, as well as the use of sources such as heating, cooling, power, water and so on.

There is a conditional classification of eco materials into three types: natural origin, natural and artificial and materials of chemical (artificial) origin.

Natural origin of bamboo. Since bamboo grows almost everywhere except of extremely cold place, and by history bamboo originated in china, its where bamboo first was recorded. Varieties of bamboo, as we see today, happened from ancient herbs about 30–40 million years ago, right up to the extinction of dinosaurs. The use of bamboo began as a food base for herbivorous animals, after that people began to use bamboo in different variations.

Given the global environmental situation on the planet, scientists are constantly looking for new ways to conserve greenery on the continents, to introduce new, and safer and more progressive production technologies. Innovations in the production of bamboo fiber helped to make a big step forward (Krawczuk 2013).

Sustainability of bamboo are in grown without pesticides or chemical fertilizers, it requires no irrigation. Bamboo rarely need to replant. It grows rapidly and can be harvested in 3–5 years, bamboo produces 35% more oxygen that an equivalent stand of trees, and sequesters carbon dioxide and is carbon neutral, it is a critical element in the balance of oxygen and carbon dioxide in the atmosphere. Bamboo is an excellent soil erosion inhibitor, it grows in a wide range of environments. Its production into fibers has lower environmental impact than other forms of fiber, especially synthetic ones. Since bamboo is a 100% natural product, energy is not expended on production, making bamboo already ready for use. For local residents of countries where bamboo is grown, this is work, in various industries providing jobs and eliminating poverty in developing countries. Cultivation of bamboo gives people building material, textiles, food and work. Economically, ecologically and socially the life of people will be in stability (Krawczuk 2013).

Bamboo fibers favorably differ from many others, because in addition to all the advantages of natural fibers have unique properties as hypoallergenic, antibacterial properties that persist even after prolonged use, high wear resistance, resistance to burnout, hygroscopic. The ability to retain ultraviolet rays and natural antistatic effect.

The good side of the bamboo that it is renewable material. Bamboo can be harvested from three to four years after ripening, in contrast to coniferous and deciduous plants that grow for 40 years. Yield at a bamboo is about 25 times above in difference from trees. Almost a thousand hectares of forests are cut down every week around the world, but using bamboo instead of wood makes it possible to cut this figure and protect forests. Also, bamboo can be used in different directions. Each piece of bamboo is used to make different products. People use bamboo roots as food and enrich the soil with mulch, from furniture to chopsticks and textiles.

Bamboo saves energy in all aspects, here bamboo provides insulation in winter and provides a cool during the summer, as the natural canopy allows you to penetrate the light, but protects the interior from the scorching sun and heat, bamboo is also a soundproof material, which is valuable in a busy city (Krawczuk 2013).

The versatility of bamboo is interesting. Since bamboo fibers are stronger than wood, the replacement of wood is quite possible for our world. At the moment, paper, charcoal, textiles, flooring, furniture are produced from bamboo, and this is only the beginning of the list. Similarly, unlike other materials such as cotton, which requires frequent spraying with chemicals, bamboo does not require any fertilizers, pesticides or herbicides in the process of growth.

10.1.5   Bamboo's Criteria for Building Industry

In the construction industry, not all types of bamboo can be used. In the construction industry, if the bamboo does not pass a certain processing of treatment, also cannot be used. Harvesting and protection from insects is the beginning for bamboo as a building material. When using bamboo, it is worth making sure that it is dried during the drying process, not dry bamboo can lead to kinks and is not subject to use in structures; Bamboo should be older than 3 years. It is recommended to use mature bamboo 4–6 years; Bamboo must be decontaminated from insects and a certain treatment; Bamboo blooms can't be used in construction, can lead to the withering of the bamboo itself; Before use it is worth checking for the presence of cracks, can lead to kinks; The term of bamboo service depends on the type, length of the bed, wall thickness, and harvesting time (Krawczuk 2013).

10.2   The Durability and Treatment of Bamboo for Building Industry

The bamboos outer hard part and the lower part of the culms are more durable the inner wall, which deteriorates faster as it is soft. Treating bamboo for long-term use is a more economical and sustainable method. With the application of proper procedures, the lifespan of bamboo can increase to 50 years (Patel 2015).

Besides making bamboo more durable there is another reason why there is supposed to be a treatment for bamboo before using it as a construction material. One of the bamboos problems is beetles and termites, they are the most common insects in bamboo. The high probability of infection by insects is due to the presence of starch and other carbohydrates in bamboo (Kumar et al. 1994).

Another common problem is mold and spores, if bamboo is not completely dry. In addition, shipping bamboo in sea freight containers to the international market, too, can affect the occurrence of spores and mold (Schröder November 16, 2012). A high moisture content in the reed makes it possible to reproduce the spore until the moisture completely evaporates. The appearance of mold can occur from one to two times (Schröder November 16, 2012).

In the fast and effective removal of stains can help commercial product Mold Armor FG502. You can also use alternative methods to combat mold. Clean the bamboo mold and white spores with a soft brush and wipe the area with a wet rag. After cleaning the bamboo, you can use lemon oil or a solution of vinegar and water for completely remove the mold. To clean the furniture from bamboo, it is recommended to apply three layers of polyurethane after removing the mold; this will ensure the preservation of furniture. Before applying the polyurethane, you need to clean the furniture with turpentine and completely dry it. If you do not do this, mold can grow under polyurethane coating and cause blisters (Schröder, November 16, 2012).

One of the case is Phyllostachys aurea and the appearance of bamboo Phyllostachys aurea is often treated with flame (heat treatment). Another case is thicker bamboo species, such as Guadua angustifolia, are supposed to be treat with boron solution.

The choice of a suitable method of treatment depends on various factors as green or dry bamboo, bamboo shape: round bamboo or half. End applications; in contact with the earth, is exposed to the atmosphere, under cover, structural / non-structural. Scale; the amount to be treated, and the available time. Since the bamboo is a green plant that can be attacked by insects, how to protect bamboo not only from insects but also from mold. Potential causes of decay; biotics (fungi / insects) and abiotic (fissures / weathering) (Schröder November 17, 2012).

Bamboo leaching is a traditional method, it is used by indigenous communities and farmers in the Asian and Latin American regions. People transported bamboo from mountain areas and jungles to cities using bamboo rafts. Transportation of bamboo over rivers usually took up to four weeks, during which time leaching occurred. The content of starch in the poles was reduced, thereby increasing the durability of the material (Schröder, November 14, 2012).

People still use the method of immersing bamboo in water, especially when bamboo needs to be transported from remote areas. However, single leaching does not guarantee long-term protection, but helps to remove starch and increases permeability for future treatment by diffusion and pressure treatment. This protects bamboo from attack by beetles and parasites (Schröder, November 14, 2012).

Store bamboo in water. Bamboo is stored in tanks with water while adding chemicals. Bamboo knots should be punctured, so water will easily penetrate into the bamboo canes. Binding the bamboo together or to separate and store them in tanks or in running water one of the main part. While the storage of bamboo in the tank necessary to change the water weekly, this will prevent the growth of bacteria that can cause unpleasant odor and discoloration of bamboo. It is necessary to use loadings for full immersion of bamboo in water. Bamboo should be immersed in water with times, after extraction, it is worth using further chemical treatment. Bamboo should be immersed for at least 3–4 weeks. Bamboo that has been leached for 3 months or more may become stained in the epidermis. This will reduce its physical and mechanical properties (Schröder, November 14, 2012).

For green bamboo, apply heat treatment, most of the moisture remains inside the canes. The dried bamboo is subjected to boron treatment. Therefore, some finished bamboo products are initially more sensitive to molding. However, if you load an ocean freight container with bamboo in a tropical environment where a long transport time takes 30 days or more and ship to an environment where the temperature is much cooler, the cargo may fall under the phenomenon known as "cargo sweat" or "container rain". Extreme temperature fluctuations can cause condensation inside the container, which causes spore growth and the appearance of mold (Schröder, November 16, 2012).

The bamboo which people use in construction cannot be picked up correctly, or is prone to improper handling, many people don't use any treatment, so the construction of their buildings can collapse in just a couple of years, because of that bamboo is still considered as a poor man's material. The countries that account for the majority of bamboo are not sufficiently informed about professional treatment. Not all methods that they use provide good results, this leads to the destruction of the construction of bamboo. For protection it is worth using chemical preservatives, they provide good protection even under adverse conditions (Schröder, November 30, 2014).

Chemicals can provide short-term and long-term protection. The selection of chemicals as protection for bamboo must be made very carefully, as this can affect the environment. Due to Schröder (November 13, 2012) there is non-fixed preservatives and preservatives of the fixing type, they help to prevent the occurrence of termites and mold and protects against the appearance of fungi.

Non-fixed type of protection consists of boron salts. Boron salts are added to water, bamboo is soaked in this water, after the water evaporates and leaves the salts inside the bamboo. This method is not toxic; it can be used for products that contact with food. Another method of non-preservatives is Boric Acid and Borax its commonly used methods of preserving bamboo and environmentally friendly. Boric Acid Borax is dissolved in water, after it can be sprayed onto the material, bamboo can be immersed or impregnated with this substance (Panday and Suneel 2007).

Preservatives of the fixing type, its type of protection consist of mixtures of different salts in a certain quantitative ratio, when interacting with each other, they become chemically fixed.

One of the preservatives methods is Copper-Chromium-Arsenic (CCA) provides protection for 50 years or more, the use of such a composition is desirable in the open air, since arsenic is a poison and a carcinogen, but the use in small quantities does not affect human health. Another method is Copper-Chrome-Boron (CCB) it is an alternative to CCA, but because of the boron degree of fixation is lower, is less effective. Zinc-Chrome it is a refractory preservative designed to protect the material from fire and provides good protection against insects. However, bamboo can absorb moisture from the air and will give a wet look during the rainy season. There is another types of preservatives as creosote, Light Organic Solvent-based Preservatives (LOSP), Trichlorophenol (TCP), Copper/Zinc soaps (Patel 2015).

Drying bamboo is a lengthy process because this material easily absorbs moistur, defined as a hygroscopic material. Humidity is 50–60%, which can also depend on the climatic conditions in which the bamboo is cut down as well as the type of bamboo. During the drying process, the diameter of the bamboo can be reduced by 10–16%, and the wall thickness by 15–17% (Schröder November 12, 2012).

The usual and well-known method of drying bamboo is "drying in the air." After chemical treatment, all bamboo is laid and stored under a canopy or under a roof. Bamboo supposed to be protected from direct contact with the groung, avoid humudity, insects and infection with fungus. One of the main part is to avoid direct sunlight, or the bamboo may crack, but if the bamboo is divided into parts, then it can be dried under the sun. There should be a circulation of air in a room with bamboo. The vertical packing gives the bamboo to dry for a shorter period and is less defeated by a fungal attack. However, it is worthwhile to follow the poles so that there is no curvature or provide a good support system. Horizontal laying is usually used for large batches of bamboo, they are laid on large sheets, and sheets can consist of plastic or glass, or use dividers. The lower bamboo batch can crack from the weight, for this, it is laid not in large layers and carefully checked. Every 15 days, bamboo should be turned in the longitudinal direction, for even drying. The method of air-drying takes from 6 to 12 weeks. This may depend on the humidity of the bamboo, the thickness (Schröder November 12, 2012).

Drying bamboo takes a lot of time, because this material easily absorbs moisture, it is called a hygroscopic material. Humidity is 50−60%, it can also depend on the climatic conditions of cutting down and the type of bamboo. During the drying process, the diameter of the bamboo can be reduced from 10 to 16%, and the wall thickness from 15 to 17% (Schröder November 12, 2012).

Post-harvesting transpiration. Transpiration is the process of movement and evaporation of 99% of water from a plant through leaves and stems. The traditional method of drying bamboo, is used by farmers and indigenous communities. The drying process takes place on a bamboo plantation, in order to avoid contact with the soil, the cut bamboo is placed on a stone and leaned against another bamboo. Attached to each other for 3−4 weeks, the bamboo begins to lose its moisture (Schröder November 12, 2012).

Kiln oven drying. The drying kiln method is suitable only for bamboo split, the process quickly dries out pieces of bamboo. However, this method is not suitable for whole bamboo, as high temperature gives cracks in the bamboo (Schröder November 12, 2012).

10.2.1 Criteria for Building Industry

In construction industry the most important part of building is joints. There is different type of joins and connection of bamboo poles. The Double Butt Bent Joint, two bamboo poles were used. The first is horizontal, the second is vertical.

The Friction–Tight Rope connection is the most common way to connect bamboo. Originally used natural materials, such as bast, strips of bamboo, rattan/hemp and palm fibre from coconut/sago, are used as a rope for connecting bamboo poles. At present, iron wire and plastic straps/ropes are also used for connection with friction (Janssen 2000). The Friction—Tight Rope connection technique can be used in almost all circumstances that may appear during the construction of a building and reduces the likelihood of a bamboo fracture. It is easier to compact because it is possible to apply epoxy or other materials as a coating. However, the choice of rope material can affect the construction, so you need to check the rope for damage from moisture, putrefaction and provide protection from insects (Janssen 2000).

Taking advantage of a secondary interlocking element, this type of connection is widely used in context with rope connections. The bolts have to transfer tractive and compressive forces. In addition, this can be seen in garden fences and furniture (Socrates 2012). This type of connection is used with a blocking element that can be made of wood and bamboo, and a cannon joint is used. This type of connection is most often seen in garden fences and furniture. To begin with, drill the vertical and horizontal bamboo beams; the vertical that is connected to the wooden cipher should have holes larger than the horizontal pole. The used pin (made of wood or bamboo) connects two bamboo poles, and then a bamboo wedge is fixed to strengthen the connection. After that, you need to cut off the protruding parts of the curtain and make the surface of the plug smooth. To strengthen the bond use an adhesive or a rope (Socrates 2012).

Positive Fitting connection creates a beautiful view of the bamboo in the design. The connection can be combined with other connections, such as plug-in technology, attachment technique. A big plus of compound is that you can reduce the cost of bamboo pillars. In addition, when small gaps are detected on the surface, the assembly can be easily sealed and fixed (Janssen 2000).

However, it is necessary to consider the connection of two bamboos, when constructing a construction, the diameters of bamboos should be different. In addition, when creating a hole on the bamboo poles, the work of experienced masters is important, as if not working correctly, the strength may decrease and may cause cracking and splitting (Regenerative Design Institute's Photostream 2007).

Special construction design is used in constructing. One of the type of bamboo connection is based on plug—in joint and friction tight rope joint (Schröder 2009). A special and durable design, very unique and easy to build. An effective and easy way of connecting does not require high skills of working with bamboo. This "sandwich" design does not require a structural frame and does not request special tools, only the bolt and nuts. The posts are connected to each other, three bamboos are used, four postures of bamboo are added from the top in the middle, five are used at the bottom, but the number of bamboo can vary depending on the characteristics of the building.

Bamboo poles and around them are usually glued and joined by blocking compounds. Interlocking connections can be made through a wood, metal anchor technique. The junction of six bamboo poles directed in different directions, made in metal anchor technology (Janssen 2000).

In blocking compounds, it is very often possible to find a connection with a wood core. For interior decoration of the wood, different parts can be used that do not violate the requirements, glue can be used to attach it to the inner surface of the bamboo. During the installation of the wooden cylinder, two slots must be present to track the cracking in the bamboo cane.

Truss construction are widely used in modern construction, mainly to cover large spans in order to reduce the consumption of materials used and to facilitate the construction, for example—in building large-span constructions, such as bridges, rafters of industrial buildings, sports facilities, as well as the construction of small lightweight building and decorative constructions: pavilions, stage constructions, awnings and podiums. People use truss construction for roof.

For architects, who are working with bamboo, they do not stop on one structure if it comes to this material. Bamboo material for which architects create a structure that may not exist; the material gives the right to implement ideas that go beyond the ordinary.

Bamboo is a material that is used for building above the ground, since there is a possibility that the bamboo will germinate underground. However, if you observe certain significant rules and do not put bamboo on the straight to the ground, then the use of bamboo as a foundation material is possible (Khatry and Mishra 2012).

Bamboo's several regulations for foundation. First is that the bamboo and soil should not come in contact there is a risk that bamboo can take root. The plinth in which the bamboo is installed should exceed 350 mm above the ground water or above the hot water line (Krawczuk 2013). For foundation diameter of bamboo should not exceed 70 mm (Krawczuk 2013). If post it exceeds three permissible meters, then connecting beams should support the column.

People in search of comfort, increasing mood and looking for a cozy atmosphere are beginning to turn to environmentally friendly materials–Bamboo was no exception. There are some types of bamboo walls, one of them is whole or halved bamboo, and it is a typical wall where the bamboo whole or cut on two is used with rope or other fastening things. There is some types of bamboo.

Split or Flattened Bamboo. Separated bamboo from the thin end. By hitting the hammer on the blade it needs to separate the reed. The splitting line will leave the center to prevent it, it is necessary to keep the expanded part of the cane by pressing it, and the smaller part to pull down, after checking the thickness of the upper and lower parts, the thin part always points downwards. Bamboo has two different sides; straight side and curved side (Guadua Bamboo 2017). By dividing the green bamboo stems into parts without breaking them into pieces, afterward, the diaphragms are removed, then unfold the bamboo and smooth it. The result is a board that is laid on the beams and fixed with nails or tied.

Bajareque. A strong and massive construction is often used in Latin America. Horizontal bamboo strips, connected together or nailed, covered with mud and stones, thereby filling the free space (Guadua Bamboo 2017).

The wattle bamboo. Thick panels of bamboo strips are intertwined around the bearing beam of bamboo. It is commonly used in India, Peru and Chile.

Roof. One of the best roofing methods of using bamboo is one that uses prefabricated bamboo farms, as well as bamboo boards. Farms are usually built in the form of a triangle, which gives the roof a completely different look than most western styles of houses. Sometimes bamboo is used with a coating of astringent materials after the construction is completed. This not only helps protect bamboo fibers, but also helps to create fireproof structures. Clay tiles can be placed in the upper part of the roof to prevent moisture penetration. When used as a building material, bamboo must be processed to prevent it from rotting or infecting insects. Although there are many different ways of preparing bamboo, the simplest and most effective is to simply allow it to air dry in an upright position. After the bamboo has been dried, a combination of chemicals is selected, for example, boric acid and borax for its processing. The entire process can take several months, but in the end, bamboo can be used in construction (Whittaker 2010).

The disadvantages of using bamboo for a roofing device. Unlike traditional wooden roofs, bamboo roofing can be quite expensive compared to other types of roofs. Bamboo is environmentally friendly and sustainable, but requires care and money investment. Although roofing bamboo can often be found on roofs in the traditionally western type of roof, it is mainly used when used with a triangular truss. If the appearance of the triangular roof does not satisfy you, you can always choose your own version. Investing in the bamboo roof structure is a big deal. For confidence in the right choice, consider all the pros and cons associated with the material, its stability and its durability in your particular area.

Bamboo reinforcement. Researchers from the BRE Centre of the University of Innovative Building Materials, in collaboration with a team from Coventry University and the University of Cambridge, are studying the use of bamboo for mass housing construction. The research team develops an understanding of the properties and structure of bamboo fibers, in order to reduce its weaknesses, while retaining its unique mechanical properties (Walker 2014).

The design concept has undergone serious changes, boldly violating the traditional notions of form. With the help of the computer-aided design system, which carries out the most complicated computing functions, it became possible to exercise much more freedom in creating the project. After the constructive solutions of the structures have overcome the established architectural patterns of the past, the original designs have undoubtedly found their vivid embodiment in modern buildings of bamboo and interior spaces of their premises.

At the World Architecture Festival in 2016, scientist Dirk Hebel presented bamboo fiber as a more solid and cheaper alternative to steel. The team, led by a German architect from the Zurich Institute of Technology (ETH Zurich), is working on a composite material based on bamboo fiber with the addition of organic resins, which in the future can change the technology of reinforced concrete construction (Walker 2014).

From the new material, you can do any forms. For example, as twigs, it can replace the reinforcing mesh without any loss in bearing capacity. At the same time, such a bamboo composite works better for stretching but weighs four times less.

Now, together with the University of Berkeley, the scientist is developing a new concrete, where instead of cement, a material based on mycelium (mycelium) would be used. The scientific laboratory from Singapore Future Cities Laboratory suggested using simple bamboo instead of reinforcement. In developing countries, steel reinforcement is very in demand. During the experiments, bamboo has outstripped the strength of many materials—including steel reinforcement. This stability is provided by the tubular structure of the bamboo stems, formed during the evolution under the influence of winds. Another undoubted advantage of bamboo is its low weight—it is easy to assemble and transport. In addition, the cultivation of bamboo as a building material will positively affect the environment—during growth, it absorbs a large amount of carbon dioxide (Walker 2014).

Bamboo as solar shading devices. External shading devices for windows are effective cooling measures, since they block both direct and indirect sunlight from outside. The shading screen is a dense weave that blocks up to 70% of all sunlight. With the proper use of mobile shading devices, you can gain advantages in the use of energy. In the daytime, the shading device is kept in the open state in winter; it does not reduce the supply of radiation energy in the room. In winter, at night, when closed, lowers the heat transfer coefficient, i.e., heat loss, and in summer it protects the room from overheating. Such screens absorb sunlight, so they should be installed from the outside of the window.

Stairs. In countries where bamboo grows, the trunks of this plant are often used to make ladders. The choice in favor of bamboo in the manufacture of stairs is completely justified: such simple designs are very convenient. Even with a long length, they are light enough, and they can be easily carried by one person. According to the planned width of the stairs, the third bamboo is cut (the trunk is cut to the appropriate lengths). Next, take a long bamboo, in the trunk of which make holes. At the same time, the diameter of the holes must be such that they include previously cut pieces of bamboo stem. Holes can be made with one (internal) or two sides of bamboo (the trunk is cut through). If the holes are made only from the inside, the transverse sections of bamboo into the trunk are pushed to the stop so that it touches the opposite side of the support. Fasten the steps in any convenient way, for example, cutting bamboo to the trunk can be glued with a suitable glue for bamboo. In order not to allow the vertical stands of bamboo to disperse under load, the trunks in the top, bottom and middle should be tightened with threaded studs. In order not to spoil the appearance of the ladder and not to make extra holes in the bamboo, the trunks are pulled together by pins, passing them inside the transverse segments.

Fence. Bamboo can be used as a frame for a balcony or a loggia, and it can also serve as a decoration for entering the house. And this is made possible by the fact that the bamboo has lush foliage, which makes it possible to create rapidly growing walls and a clever background for landscape compositions.

Bamboo on construction sites. Builders work with some tools and devices made of bamboo. These are ladders, scaffolding, and supports (for example, for horizontal formwork). Because of the relatively high cost of bamboo in the temperate countries in this capacity, it can be used where a small amount of material is required, for example in short ladders in the backyard (Chan et al. 1998). Such a ladder with a sufficient diameter of the trunks will come out strong and light, which is important in its frequent transfers. You can also make an overhead staircase on the roof during its repair. It will also be easy to move it.

Other building elements. It is possible to color this natural material in different colors—from light to dark. The surface of the canvas is varnished, giving them a shine. These products, thanks to processing by modern technology, perfectly withstand temperature changes, high humidity and exposure to sunlight. Modern industry allows you to make of banal bamboo stylish and high-quality furniture, which look great and do not require special care.

Another advantage of such products is their low price. As example the original door can afford almost every buyer, buying a little affordable exotics. Do not think that such lightweight and airy structures have a low resistance to burglary. In fact, bamboo is not inferior in strength to metal, and it is very difficult to crack such a door structure. "Sheathing" is made of strips of glued bamboo, laid on the fabric base, which gives them high wear resistance. The basis for such door panels is made of strong steel, treated with an anti-corrosion compound. This allows similar products to withstand heavy loads and exposure to adverse weather conditions.

10.3 Material and Methods

Using MCDM methods, working bamboo types and depending on the best for the worse is feasible.

Collected data were entered into the PROMETHEE application to calculate it (Ozsahin et. al. 2019a, b) Fuzzy initial data is used for decision-making problems associated with subjectivity, insufficient or partial information, quality information, and to give a rough estimate.

Table 10.1 shows a linguistic scale for the assessment of criteria, showing a triangular fuzzy scale with priority ranking (Uzun et al. 2019).

Table 10.1 Linguistic Scale for Importance

Linguistic scale for assessment	Triangular fuzzy scale	Significance of criteria ratings
Very high (VH)	(0.75, 1, 1)	Height average (m), Wall Thickness av. (cm)
Important (H)	(0.50, 0.75, 1)	Diameter av. (cm)
Medium (M)	(0.25, 0.50, 0.75)	Construction Grade from 1 to 10
Low (L)	(0, 0.25, 0.50)	Hardiness (°C)

The dimensions shown in the table are based on the green state of bamboo (Plan-tUse, Flora of China, Guadua Bamboo, PFAF, Useful Tropical Plants), as after careful treatment and drying of the bamboo, the dimension's change. The change in size depends on for the type of treatment that is chosen for the bamboo. For frequent bamboo, sizes change by 10–15% after treatment (Tables 10.2 and 10.3).

The bamboo that people use in construction is not always selected properly or is prone to improper handling, and many people do not use any treatment, so buildings constructed using this material can collapse in just a couple of years. The selection of the type of bamboo depends on the kind of building the architects plan to construct. Table 10.4 showing the ranking of the material used it is graded from the most to the least useful.

10.4 Results and Discussion

10.4.1 Results

The paper shows the work of combination of the MCDM methods with fuzzy logic to determine which species of bamboo is better to use in construction industry. In the Table 10.2 the H (m), D (cm), Wall T. (cm), Internode l. av. (cm), Const. (1–10) this is one of the main criteria in the selection of the right type of bamboo for construction and design. However, if changes are made to the criterion part of the table, the result also changes. The results show that the first three in the ranking are the most suitable type of bamboo: # 1–Bambusa blumeana, # 2–Phyllostachys edulis and # 3–Dendrocalamus giganteus. The main advantage is its user-friendliness arising from linguistic evaluations and it also takes into account the uncertainty or fuzziness inherent in a particular topic. When it comes to bamboo, you cannot accurately say its dimensions. As many buildings can be built in different ways and a certain bamboo will be selected for each element of the building. Bamboo is difficult to make the same for large batches, so builders and engineers should select bamboo separately for each element of the building. For architects who work on building bamboo buildings the provided table will be useful since it indicates the size of the bamboo and where it can be used.

With the right approach and processing of this material a house can be build using bamboo from the foundation to the roof. Bamboo releases 35% more oxygen than coniferous trees while also absorbing carbon dioxide. Since it is a 100% natural product. Energy is not expended on production, meaning that bamboo is ready for use. Bamboo can play an important role in the future of mankind as the human need for wood can be replaced with bamboo thus preventing frequent deforestation due to the growing demand for raw materials for housing and construction.

Table 10.2 Visual PROMETHEE application for the bamboo types

Scientific Name	Ha. (°C)	H. (m)	D. (cm)	Wall T. (cm)	Internode l. (cm)	Const (1–10)
Arundinaria alpina	−4	20	12	3	25	4
Bambusa bambos	−1	30	15	3	30	9
Bambusa balcooa	−4	18	10	2	30	10
Bambusa blumeana	−1	20	20	3	45	10
Bambusa dolichoclada	−3	13	8	1.5	38	7
Bambusa multiplex	−8	5	3	1.5	30	6
Bambusa nutans	9	12	7	2	30	6
Bambusa pallida	3	17	7	1	50	5
Bambusa polymorpha	0	20	12	1.5	50	6
Bambusa textilis	−4	9	4	0,3	60	6
Bambusa tulda	9	15	10	2	55	7
Bambusa tuldoides	−7	12	4	4	33	5
Bambusa vulgaris	−2	20	8	1	30	7
Cephalostachyum pergracile	−1	20	6	1	38	4
Dendrocalamus asper	-4	25	14	2	35	10
Dendrocalamus barbatus	15	16	18	1	30	5
Dendrocalamus brandisii	−3	27	17	3.5	34	9
Dendrocalamus giganteus	−1	30	22	2	40	6
Dendrocalamus hamiltonii	−1	15	15	1.5	40	7
Dendrocalamus hookerii	−1	18	13	2	45	7
Dendrocalamus latiflorus	−4	19	15	2	50	8
Dendrocalamus longispathus	15	15	9	1	33	9
Dendrocalamus membranaceus	−4	22	8	0.8	32	7

(continued)

Table 10.2   (continued)

Scientific Name	Ha. (°C)	H. (m)	D. (cm)	Wall T. (cm)	Internode l. (cm)	Const (1–10)
Dendrocalamus merrillianus	15	18	8	2.5	18	10
Dendrocalamus sikkimensis	−2	18	13	2	38	10
Dendrocalamus strictus	−5	15	6	4	38	7
Gigantochloa albosiliata	10	12	5	1	35	7
Gigantochloa apus	−2	15	9	1	45	8
Gigantochloa levis	−1	17	10	1	25	6
Gigantochloa macrostachya	−1	13	1	0.3	60	6
Gigantochloa verticillata	5	15	9	2	50	9
Guadua angustifolia	5	23	15	3	20	10
Melocanna baccifera	−1	15	7	1	40	6
Oxytenanthera nigrociliata	−1	18	45	4.5	28	9
Phyllostachys aurea	−15	8	3	1	9	5
Phyllostachys bambusoides	−15	20	14	5	25	10
Phyllostachys edulis	−4	25	18	1	40	10
Pseudostachyum polymorphum	1	28	3	1	20	8
Schizostachyum hainanense	1	15	3	1.5	55	7
Thyrsostachys oliverii	5	20	6	1	50	10

Hardiness (°C)—The temperature that bamboo tolerate.
H (m)—Average height of the bamboo.
D (cm)—Average diameter.
Wall T. (cm)—Average wall thickness.
Internode l. av. (cm)—Average internode length.
Const. (1–10)—Grade of construction uses from 1 to 10.

Table 10.3 Grade of construction uses from 1 to 10

Construction uses of bamboo	Rate
Basketry, handicrafts	1
Handicrafts furniture, basketry, floor strips	2
Board, floor strips, poles, handicraft furniture	3
Furniture, poles, rafters, fences, light construction, walling mats, shingles	4
Framing, walls, floor, sheathing, farm equipment, punting poles, scaffolding, light construction, furniture, farm implements, basketry, and handicraft furniture	5
Light construction, walls, furniture, woven, landscaping, sheathing matting, fences, boards, parquet, rural housing, frames of thatched roofs, floor strips, sheathing, troughs, pipes	6
The building material for houses, frame construction, furniture, scaffolding, wall, trough, particularly roofing, floor, concrete formwork–boards, sheathing-strips, board, furniture, matting roof-shingles	7
Building construction, agricultural implements, furniture, for structural timber (of medium quality) in houses and temporary construction, roofing, scaffolding, walls, fences for weaving into paneling, sheathing	8
Construction material, reinforcement, concrete formwork and framework, handicrafts, furniture, mats and containers, floor and sheathing–strips, wall, roof, scaffolding	9
Concrete reinforcements, alternative to wood for the production of laminated and agglomerate wood (columns, beams, girders, planks, panels, etc.) parquets, furniture, structural timber for heavy construction such as houses and bridges, furniture, strip floor, sheathing strips, concrete framework and formwork, roof, lashing, wall, sheathing, scaffolding	10

10.4.2 Discussion

Based on the data collected, this work showed that there are not many types of bamboo that we can use as building material, however, there are quite a few different types of bamboo to understand that we need to put it in the ranking method. For multi-criteria analysis although it is comparable with other multicriteria decision making theories, but the most advanced numerical and nonnumeric data making theory method is Fuzzy PROMETHEE. The top ranked bamboos are Bambusa blumeana. Phyllostachys edulis. Dendrocalamus giganteus. While the lowest ranked are Bambusa tuldoides. Bambusa multiplex, and Phyllostachys aurea. Analysis of the table showing the type criteria indicates that Bambusa blumeana has good criteria, but not the best. Furthermore, Phyllostachys aurea can survive at −15 °C and the criteria are not the worst. Not every type of bamboo is suitable for reinforcement or construction. As the properties of each type should be taken into account and experiments should be conducted with different types of bamboo.

Bamboo, despite its light weight, is very flexible and hardy. There are minuses, it is cylindrical and hollow inside. Bamboo does not always grow straight, and working with a curved plant is quite difficult. With the right approach and processing of this material, you can build a house from the foundation to the roof. One of the most

Table 10.4 Ranking of bamboo as a construction material with MCDM method with fuzzy logic

Rank	Scientific name	Phi	Phi+	Phi−
1	Bambusa blumeana	0.3549	0.4538	0.0990
2	Phyllostachys edulis	0.3149	0.4454	0.1305
3	Dendrocalamus giganteus	0.3090	0.4431	0.1342
4	Dendrocalamus brandisii	0.2765	0.4161	0.1395
5	Thyrsostachys oliverii	0.2327	0.3837	0.1510
6	Bambusa bambos	0.2310	0.3872	0.1562
7	Dendrocalamus asper	0.2253	0.3844	0.1591
8	Dendrocalamus latiflorus	0.2181	0.3448	0.1266
9	Bambusa polymorpha	0.1732	0.3260	0.1527
10	Guadua angustifolia	0.1707	0.3892	0.2185
11	Bambusa tulda	0.1675	0.3359	0.1683
12	Gigantochloa verticillata	0.1581	0.3270	0.1689
13	Oxytenanthera nigrociliata	0.1562	0.3532	0.1970
14	Dendrocalamus sikkimensis	0.1444	0.3064	0.1621
15	Dendrocalamus hookerii	0.1347	0.2942	0.1595
16	Dendrocalamus hamiltonii	0.0378	0.2456	0.2077
17	Bambusa pallida	0.0259	0.2634	0.2375
18	Phyllostachys bambusoides	0.0251	0.3230	0.2978
19	Dendrocalamus barbatus	0.0082	0.2763	0.2681
20	Dendrocalamus latiflorus	0.0071	0.2483	0.2411
21	Gigantochloa apus	0.0025	0.2255	0.2230
22	Schizostachyum hainanense	−0.0092	0.2420	0.2512
23	Dendrocalamus merrillianus	−0.0199	0.2611	0.2810
24	Bambusa balcooa	−0.0358	0.2126	0.2484
25	Dendrocalamus membranaceus	−0.0710	0.1953	0.2663
26	Pseudostachyum polymorphum	−0.0922	0.2376	0.3299
27	Bambusa vulgaris	−0.1082	0.1607	0.2689
28	Cephalostachyum pergracile	−0.1104	0.1963	0.3067
29	Gigantochloa macrostachya	−0.1156	0.2184	0.3340
30	Melocanna baccifera	−0.1201	0.1634	0.2835
31	Dendrocalamus strictus	−0.1201	0.1763	0.2964
32	Bambusa dolichoclada	−0.1542	0.1450	0.2992
33	Gigantochloa albosiliata	−0.1591	0.1790	0.3381
34	Arundinaria alpina	−0.1650	0.1886	0.3536
35	Bambusa textilis	−0.1752	0.1995	0.3747
36	Gigantochloa levis	−0.2077	0.1236	0.3313

(continued)

Table 10.4 (continued)

Rank	Scientific name	Phi	Phi+	Phi−
37	Bambusa nutans	−0.2079	0.1521	0.3601
38	Bambusa tuldoides	−0.3529	0.1075	0.4604
39	Bambusa multiplex	−0.4992	0.0419	0.5411
40	Phyllostachys aurea	−0.6501	0.0035	0.6536

basic and difficult parts is the bamboo joints, they are not typical and the methods of joining the tree or any other material will not fit in any way to the bamboo. For the bamboo mastered their own special techniques, and many architects will improve them making the design of the building exotic. Familiar homes, we begin to build with the laying of the foundation, the construction of columns, walls and finish the roof. In the case of bamboo, the foundation can flow into the supporting column, which is a wall and this structure can end as a dome-roof. Do not forget that it is durable, now scientists have studied bamboo and the mix of cement with bamboo gives a double strength, so the benefit of bamboo in subsequent years is that it can replace iron and steel in the reinforcement of buildings.

10.5 Conclusion

Bamboo as a building material is exactly one of the worthiest materials that does not harm the environment. Surprisingly it was that this herb can grow over a day by several meters and after three or four years it can be cut down and used as a material for construction. People have used bamboo for centuries in their lives. Today, bamboo is used in many areas of activity, from food and reaching construction, textiles and high technologies.

Having considered bamboo from the outside and inside, it seems that this is an ecological product that people still do not know how to use. Protection from pests always comes first in people who build bamboo. For local residents of countries where bamboo is grown, this is work, in various industries providing jobs and eliminating poverty in developing countries. Cultivation of bamboo gives people building material, textiles, food and work. Economically, ecologically and socially the life of people will be in stability.

Bamboo, a unique material from which you can build houses, schools, restaurants, church pavilions, bridges and a list on this just begins. Properties of bamboo can be both a finishing material and the basis of a building's construction.

In the future, the bamboo sector should be brought to the forefront as a sustainable material; bamboo is not well represented to the public, as it has just begun to gain the attention of architects and investors. With a good contribution to bamboo, its use can significantly increase and many countries in Europe and nearby countries in the tropical and subtropical regions can learn more about bamboo. Bamboo can play an important role in the future of mankind, the human need for wood can be replaced with bamboo, while wood will be protected from frequent deforestation, due to the growing demand for raw materials for housing and construction. In addition, a big plus of bamboo, this is the secondary use of the material.

References

amaZulu Inc (2014) The many bamboo uses throughout history to today." www.amazuluinc.com/2014/03/03/the-many-bamboo-uses-throughout-history-to-today/

Chan SL, Wong KW, So YS, Pon SW (1998) Empirical design and structural performance of bamboo scaffolding. In: Proceedings of the symposium on bamboo and metals Scaffoldings, The Hong Kong Institution of Engineers

Flora of China (n.d.) Retrieved March 26, 2020 from https://www.efloras.org/flora_page.aspx?flora_id=3

Community Architects Network (CAN) (May 2013) In: Bamboo construction source book

Guadua Bamboo (2007–2017) Crushed bamboo mats. Retrieved March 2 2018 from https://www.guaduabamboo.com/crushed-bamboo/

Guadua Bamboo (n.d.) Retrieved March 26, 2020 from https://www.guaduabamboo.com/

Hands-on-chinese style bamboo furniture (2003) Manual on bamboo furniture making. INBAR. Retrieved April 10, 2018 from https://www.inbar.int/wp-content/uploads/2013/08/Bamboo-processing-Furniture-Manual-PDF.pdf?7c424b. Accessed 27 Feb 2014

Introduction to Multiple Attribute Decision-making (MADM) Methods (n.d.) In: Springer series in advanced manufacturing decision making in the manufacturing environment, pp 27–41. https://doi.org/10.1007/978-1-84628-819-7_3

Jamatia S (2015) Livelihood of the Bamboo base: challenges and Opportunities. Accessed from https://www.academia.edu/3794654/Livelihood_of_the_Bamboo_base_Challenges_and_Opportunities

Janssen JJA (2000) Designing and building with bamboo. In: Arun K (ed) International network for bamboo and rattan, Technical University of Eindhoven, the Netherlands

Khatry R, Mishra DP (2012) Finite element analysis of bamboo column along with steel socket joint under loading condition. Int J Appl Eng Res 7(11):

Krawczuk K (2013) Bamboo as sustainable material for future building industry. KEA-Københavns Erhvervsakademi

Kumar S, Shukla K, Dev T, Dobriyal P (1994) In: Bamboo preservation techniques: a review. International Network for Bamboo and Rattan and Indian Council of Forestry Research Education

Multicriteria Personnel Selection by the Modified Fuzzy VIKOR Method (2015) Sci World J 1–16. https://doi.org/10.1155/2015/612767

Ozsahin DU et al (2019) Evaluation and simulation of colon cancer treatment techniques with fuzzy PROMETHEE. In: Advances in science and engineering technology international conferences (ASET). https://doi.org/10.1109/icaset.2019.8714509

Ozsahin I, Sharif T, Ozsahin DU, Uzun B (2019) Evaluation of solid-state detectors in medical imaging with fuzzy PROMETHEE. J Instrum 14(01). https://doi.org/10.1088/1748-0221/14/01/c01019

Panday S (2007) Preservation of Bamboo. NMBA, TIFAC, DST (GoI), New Delhi

Patel A (2015) Bamboo structures. School of Mechanical, Aerospace and Civil Engineering

PFAF (n.d.) Retrieved March 26, 2020, from https://pfaf.org/user/Default.aspx

PlantUse (n.d.) Retrieved September 26, 2020, from https://uses.plantnet-project.org/en/Main_Page

Regenerative Design Institute's Photostream (July 23, 2007) Bamboo joint. Retrieved May 20, 2018, from https://www.flickr.com/photos/regenerativedesign/879623512/

Schröder S (2014). Building Walls with Crushed Bamboo. Retrieved April 20, 2018 from https://www.guaduabamboo.com/working-with-bamboo/building-walls-with-crushed-bamboo

Schröder S (May 18, 2009) Bamboo joints and joinery techniques. Retrieved April 21, 2018 from https://www.guaduabamboo.com/working-with-bamboo/joining-bamboo

Schröder, S. (Nov 12, 2012) Drying bamboo poles. Retrieved April 18, 2018, from https://www.guaduabamboo.com/preservation/drying-bamboo-poles

Schröder S (Nov 13, 2012) Chemical Bamboo Preservation. Retrieved April 23, 2018, from https://www.guaduabamboo.com/preservation/chemical-bamboo-preservation

Schröder S (Nov 14, 2012) Leaching Bamboo Retrieved April 15, 2018 from https://www.guaduabamboo.com/preservation/leaching-bamboo

Schröder S (Nov 16, 2012) How to remove Bamboo mold. Retrieved April 19, 2018 from https://www.guaduabamboo.com/preservation/how-to-remove-bamboo-mold

Schröder S (Nov 17, 2012) Bamboo insect infestation. Retrieved April 18, 2018 from https://www.guaduabamboo.com/preservation/bamboo-insect-infestation

Schröder S (Nov 30, 2014). Durability of Bamboo. Retrieved April 16, 2018 from https://www.guaduabamboo.com/preservation/durability-of-bamboo

Socrates N (2012) In: Bamboo construction

Useful Tropical Plants (n.d.) Retrieved March 26, 2020, from https://tropical.theferns.info/

Uzun B, Yildirim FS, Sayan M, Sanlidag T, Ozsahin DU (2019) The use of fuzzy PROMETHEE technique in antiretroviral combination decision in pediatric HIV treatments. In: 2019 Advances in science and engineering technology international conferences (ASET). https://doi.org/10.1109/icaset.2019.8714389

Walker C (2014) Bamboo: a viable alternative to steel reinforcement? (8 June 2014) www.archdaily.com/513736/bamboo-a-viable-alternative-to-steel-reinforcement.

Chapter 11
Ranking of Natural Wastewater Treatment Techniques by Multi-criteria Decision Making (MCDM) Methods

Tagesse Gichamo, Hüseyin Gökçekuş, Dilber Uzun Ozsahin, Gebre Gelete, and Berna Uzun

Abstract Wastewater treatment is the method for the elimination of pollutants from effluent to release the effluent to the environment that has a pollutant concentration at an acceptable limit or to directly reuse the water. Recently, the world is demanding the wastewater treatment systems which is affordable economically and technically, carbon-free and ecological friend. In this study, the efficiency of five several natural wastewater treatment approaches namely; waste stabilization pond, soil filter, constructed wetland, use of aquatic plants, and reusing of wastewater for irrigation were compared using VIKOR. For the evaluation, different criteria were used such as land requirement, pollutant removal capacity, health risk, capital cost, hydrogeological risk, maintenance cost, subject to seasonal variability, and ecological benefits. The ranking result indicated that stabilization pond, constructed wetland,

---

T. Gichamo (✉) · H. Gökçekuş · G. Gelete
Faculty of Civil and Environmental Engineering, Near East University, Nicosia, Turkish Republic of Northern Cyprus, Turkey
e-mail: tagesseg@gmail.com

H. Gökçekuş
e-mail: huseyin.gokcekus@neu.edu.tr

T. Gichamo · G. Gelete
College of Environmental Science, Arsi University, 193 Asela, Ethiopia

D. Uzun Ozsahin · B. Uzun
DESAM Institute, Near East University, Nicosia, Turkish Republic of Northern Cyprus, Turkey
e-mail: dilber.uzunozsahin@neu.edu.tr

D. Uzun Ozsahin
Department of Biomedical Engineering, Near East University, Nicosia, Turkish Republic of Northern Cyprus, Turkey

Medical Diagnostic Imaging Department, College of Health Science, University of Sharjah, Sharjah, United Arab Emirates

B. Uzun
Department of Mathematics, Near East University, Nicosia, Turkish Republic of Northern Cyprus, Turkey

soil filter, use of aquatic plants, and reuse of wastewater for irrigation ranked from best to worst respectively.

Keywords VIKOR · Wastewater · Treatment · Natural

11.1 Introduction

The number of population in the world has almost doubled in the last fifty years, the consumption pattern in association with climate change are putting significant pressure generally on natural resources and specifically on water resources (Gichamo and Gökçekuş 2019; Sadr et al. 2018) and this resulted on a massive generation of wastewater. In recent decades, fast industrialization and manufacturing aggravated critical complications by directly releasing industrial wastes such as heavy metallic ions, dyes, and organic food wastes to the aquatic environmental (Gichamo et al. 2020). A globally escalating wastewater release has forced the institutions to treat wastewater and improve the quality before releasing it to the environment. Freshwater scarcity and climate changes enforce the world to look at alternative sources of water for consumptive demands.

Treating and reuse of wastewater is useful to fairly allocate limited water, and to reduce environmental pollution (Ayyıldız and Özçelik 2018). In the last few decades, quality standards have been increased for treated wastewater effluents that subsequently increased power demands, technology advancement, and overall costs. Conventional wastewater treatment infrastructures were commonly used and natural wastewater treatment facilities also emerged as alternative techniques. The expensiveness of the conventional wastewater treatment methods created economic pressure even on developed nations therefore, innovative, cheap, environmentally friendly methods are needed to manipulate wastewater (Tsaur et al. 2002). The selection of appropriate wastewater treatment (WWT) methods for environmental safety, health, and ecological sustainability depends on factors such as cost, energy consumption, carbon emission, and technical requirements (Bottero et al. 2011; Sun et al. 2020). Recently, different countries around the world have turned the face towards technically and economically feasible wastewater treatment solutions like natural wastewater treatment approaches, and a combination of natural and conventional WWT methods.

Natural techniques of WWT predominately rely on natural techniques to treat wastewater and it does not depend on external energy sources for major processing, however, it may require pipes and pumps for conveyance (Crites et al. 2014a, b). Natural WWT methods such as constructed wetlands, stabilization pond, reuse of wastewater for irrigation, and aquatic plants are integrated WWT options that could reduce waste load and at the same time may maintain ecological and environmentally sustainable use, health and well-being (Mitsch 2012).

The cost that natural WWT techniques required is cheaper as compared with conventional WWT methods. Treating a gallon of wastewater by natural method costs

0.6 USD however, treating the same amount of wastewater by conventional methods costs 6.5 USD (Hunter et al. 2018) with acceptable pollutants removal efficiency. The natural wastewater treatment methods have been used both on developing and developed nations because it is an important option for climate change adaptation in addition to good performance (Crites et al. 2014a). The ideal WWT strategies ought to accomplish the objectives of less treatment cost, least release of water pollutants, high maintenance capability, and providing water for reuse (Ouyang et al. 2015). Natural WWT technologies, for example, soil filters, constructed treatment wetlands, aquatic plants, waste SP, and the reuse of wastewater for irrigation engaged in protecting the nature and environment.

The natural WWT techniques comprehensively manage the sustainability of environmental components and overseeing explicit parts of wastewater while treating wastewater (Rozkosny et al. 2014). The methods achieved best results in removing most pollutants especially, organic wastes, heavy metals, total nitrogen, and total phosphorus, and produces the effluents with less chemical and biochemical oxygen demand (Qian et al. 2007). The most remarkable benefits of natural WWT methods are that it can comply with environmental standards and an excellent resilience ability despite large size land requirements and seasonal viability of performance (Ho et al. 2018). In most natural wastewater treatment methods, soil, vegetation, and microorganisms play a major pollutant cleaning roles. Natural WWT requires the large size of land for the treatment processes and to be effective, it is recommended for small villages and cities with the population size of up to 2000 households, however; it could be used at large metropolitan cities through decentralized operations (Gichamo et al. 2020; Ouyang et al. 2015). The treatment performance of different natural WWT techniques is not similar and it is not easy to quantify since it depends on different external and internal factors. In this approach, the best appropriate wastewater treatment selection decisions could be supported by Multi-Criteria Decision Making (MCDM) methods. Numerous MCDM approaches such as analytical hierarchy processes (AHP), The fuzzy-preference ranking organization method for enrichment evaluations (Fuzzy PROMOTHEE), the technique for order of preference by similarity to ideal solution (TOPSIS) and analytical network processes (ANP) has been used for preference and ranking of wastewater treatment methods (Gichamo et al. 2020; Kim et al. 2013; Ouyang et al. 2015) Therefore, the MCDM approach was confirmed as an accurate method to identify the best appropriate and effective techniques of natural WWT. VlseKriterijumska Optimizcija I Kaompromisno Resenje (VIKOR) is among powerful MCDM tools for decision making which focuses on setting ranks for different alternatives having multiple evaluation criteria. VIKOR was developed for multiple-criteria optimization of complex systems in meaningful solutions and sets rank for alternatives based on the principle of 'closeness to the ideal solution'. In this assumption, the solution closer to the ideal solution is the optimum solution or the best solution among the alternatives under comparisons. To the best of author's knowledge; this method has not applied to select appropriate natural wastewater treatment methods. Hence, this study aimed proposing VIKOR as an approach for ranking natural wastewater treatment methods using the given selection criteria.

11.2   Natural WWT Methods

11.2.1   Waste Stabilization Pond

The waste stabilization pond is a shallow earthen basin that can store wastewater and form a sequence of aerobic, anaerobic, and facultative ponds. Constructed wetland is planted with a combination of floating, submerged, and emergent vegetation (Dubey and Sahu 2014). The combination of the vegetation could regulate and preserve the physical, chemical, and biological properties of wastewater. Stabilization pond is very suitable for warm-climate and developing nations because of the availability of sunlight and high temperature, land availability, and simplicity of operation. Waste stabilization pond is appropriate in areas where land is cheap, its operation and maintenance cost is very competitive and it could produce high-quality effluent with less cost.

11.2.2   Constructed Wetland

Constructed wetlands are artificial marshy areas constructed by earthmoving, land leveling, and grading which contains wetland vegetation as filtration structures such as cattail, canary grass, and reeds. The vegetation has certain filtration structures and different paths of wastewater flow that could reduce pollutants as the water flows from path to path. When the sewage passed through the media, the impurities remain behind, and more pure water passed over, where composite chemical, biological and physical wastewater treatments are taken place (Rozkosny et al. 2014). Unlike natural marshland, the constructed wetland has a good hydraulic regime which could improve effluent quality and increase the treatment efficiency of the system (Crites et al. 2014b). The most common types of constructed wetland treatment methods are sub-surface flow, free surface water flow, and vertical flow types. In the sub-surface flow type, the water level should be below the top of the wetland, and wastewater allowed passing through permeable media to increase purification of water. Vertical flow wetland consists of sand and gravel as filters; wastewater is sprayed over the filters and facilitated pure water to pass vertically. In vertical flow wetlands, the vertical flow of water through filtration substrate is very slow, hence it is nutrient removal efficiency is high (Dubey and Sahu 2014) however, temperature fluctuations could decrease the efficiency. In the constructed wetland treatment system, the plant root zone symbiotic relationship between plants and microorganisms plays a major role in removing pollutants. The oxygen released during respiration of leaves is very crucial for metabolic processes of microorganisms thus speed up absorbing essential nutrients and breaking-down of contaminants (Kuschk et al. 2003). In the symbiotic relation between plants and microorganisms, wastewater could be treated through the uptake of organic and inorganic compounds, the release of carbon by vegetation, and breakdown of pollutants. Constructed wetland, well-vegetated with water hyacinth could efficiently remove phosphorous, nitrogen, suspended solids, sulfur, and heavy metals (Laitinen et al. 2017).

11.2.3   Use of Aquatic Plants

In this natural wastewater treatment method, vegetation grown on the aquatic environment is used as a wastewater treatment technique. In this treatment method, floating or submerged plant species such as reed, water hyacinth, duckweed, and pondweed are used to absorb and remediate heavy metals and organic nutrients; hence it could increase treatment performance of the method (Qian et al. 2007). The role of aquatic plants in wastewater treatment is mainly bacterial transformation and physical, chemical processes such as precipitation, adsorption, and sedimentation (Gersberg et al. 1986). Aquatic plants can transport oxygen from shoot to root zone, hence it will oxidize the environment which could stimulate decay of organic wastes and growth of nitrifying micro-organisms. The increased population of nitrifying micro-organisms can easily convert ammonia into nitrate. The newly formed nitrate will be removed from the system through denitrification.

11.2.4   Reuse of Wastewater for Irrigation

Reuse of wastewater for agriculture significantly improve the quality of wastewater as it applied to the soil, the organic pollutants in the wastewater are consumed as nutrients by crops (Akber et al. 2008). Recycling of wastewater for irrigated agriculture is useful for crop producers in the conservation of freshwater, increases the nutrient content of the soil, and reduces surface and groundwater pollution since it is sifted through the soil when irrigated. Raw wastewater is applied to household irrigations in uncontrolled sectors that can help farmers who cannot afford fresh water. Farmers opt to utilize wastewater to save costs paid to buy fertilizers (Corcoran et al. 2010) because the wastewater is reaching with essential nutrients, for example, 50 mg/L of nitrogen, 10 mg/L phosphorus and 30 mg/L of potassium.

11.2.5   Soil Filters

Soil filter is a cheap natural wastewater treatment method that could improve the quality of wastewater while it seeps through the soil (Akber et al. 2008). In this wastewater treatment method, soil plays natural purifying roles (Rozkosny et al. 2014) which is designed to allow water flow through the media both vertically and horizontally. Soil filters are cylindrical or prismatic in shape which is made up of concrete mixture or plastic materials and the soil is excavated by waterproof materials located on the trench. In this treatment technique, water purification processes are operated by allowing wastewater to infiltrate through the soil profile, hence, it could effectively remove biochemical oxygen demand and organic carbon (Ouyang et al. 2015).

11.3 Methodology

11.3.1 VlseKriterijumska Optimizcija I Kaompromisno Resenje (VIKOR)

Visekntenjumska Optimizacija I Kompromisno Resenje (VIKOR) is the Multi-criteria Optimization tool that could compromise the solution to the ideal solution. The idea of compromising the solution was first introduced by (Yu 1973) and the theory and applications were developed and published in 1980 (Duckstein and Opricovic 1980). The VIKOR method is a multi-criteria decision making analysis, and like any other method of analysis, there are some steps to follow to solve problems. This method can be considered linear normalization. The basic concept here is defining the positive and negative ideal points in the solution space. VIKOR is an effective decision-making tool especially when the decision-maker does not have an ability or clear evidence to prefer among the given alternatives at the beginning of system design. Unlike the TOPSIS method, which will also be discussed in this issue determines a solution with the shortest distance from the ideal solution and the farthest distance from the negative ideal solution the VIKOR method does not consider the relative importance of the distances. The primary assumption behind VIKOR is each alternative should be evaluated based on the given criteria and ranked based on relative closeness to the ideal solution and the measure of final rank by multiple criteria is evaluated by L_p metrics presented as follows.

$$\left\{ \sum_{i=1}^{n} [w_i (f_i^* - f_{ij})/f^* - f_i^-]^p \right\}^{1/p} \quad 1 \le p \le \infty j = 1, 2 \ldots m. \quad (11.1)$$

where m denotes the alternatives, (f_j^*) and (f_j^-) are the and worst criterion values respectively, for all alternatives and f_{ij} is the criterion value for the given alternative. The VIKOR compromise ranking algorithm has the following steps:

Step 1. Establish the decision matrix (Table 11.1)

Step 2. Calculate the best (f_j^*) and the worst (f_j^-) values of each criterion

In this step the best (f_j^*) and the worst (f_j^-) values should be calculated for each evaluation factor. If the aim of the criterion-j is defined as maximum, and the best value would be calculated by the formula (11.2):

$$f_j^* = \max_i f_{ij} \quad (11.2)$$

Table 11.1 General form of a decision matrix for the use of VIKOR technique

Alternatives/criteria	Criterion 1	Criterion 2	...	Criterion n
Alternative 1	f_{11}	f_{12}	...	f_{1n}
Alternative 2	f_{21}	f_{22}	...	f_{2n}
...	...	...	...	...
Alternative m	f_{m1}	f_{m2}	...	f_{mn}

If the aim of the criterion-j is defined as a minimum, and the best value will be calculated by the formula (11.3):

$$f_j^* = \min_i f_{ij} \tag{11.3}$$

Step 3. Calculate the Utility (S_i) and Regret (R_i) measures as shown in Eqs. (11.4) and (11.5), where w_j denotes the weights of the criteria, which represents the relative importance degrees for each criterion:

$$S_i = \sum_{j=1}^{n} w_j \left[\frac{f_j^* - f(ij)}{f_j^* - f_j^-} \right] \tag{11.4}$$

$$R_i = \max_i \left(w_j \left[\frac{f_j^* - f(ij)}{f_j^* - f_j^-} \right] \right) \tag{11.5}$$

The result achieved by $min_j S_j$ is the maximum group service (rule of majority), and the result obtained by $mini_j R_j$ is the minimum single regret of the 'opponent'.

Step 4. Calculating the index of VIKOR of Q_i:

Q_i can be calculated with the relationship shown in Eq. (11.6):

$$Q_i = v \left[\frac{S_i - min(S_i)}{max(S_i) - min(S_i)} \right] + (1 - v) \left[\frac{R_i - min(R_i)}{max(R_i) - min(R_i)} \right] \tag{11.6}$$

Here v can take any value from 0 to 1 and indicates the weights of the strategy that provides the maximum group utility, while $(1 - v)$ shows the weight of the individual regret. The value of v can be considered as 0.5.

Step 5. Rank the alternatives based on the Q_i, R_i and S_i values in decreasing order. This will provide three lists to the decision-maker. The alternative with the lowest VIKOR value is the best value and an alternative with the minimum Q_i provides the best solution (A') if it satisfies the following conditions:

Condition 1: (Acceptable advantage)

This condition states that there is a difference between the best and the closest to the best option.

$$Q(A'') - Q(A') \geq DQ \tag{11.7}$$

where $Q(A'')$ has the second minimum values of Q_i and $DQ = 1/(m-1)$ where m denotes the number of the alternatives then A'

Condition 2: (Acceptable stability)

A' must have the best value/s of the R_i and/or S_i amongst the other alternatives.

In this method, if one of the conditions is not satisfied, then the compromise solutions set can be proposed as below:

- If only the second condition is not satisfied; A' and A''
- If the first condition is not satisfied; $A', A'', \ldots A^M$ where M determined as the maximum decision points satisfy the condition $Q(A^M) - Q(A') < DQ$.

For the ranking of best wastewater treatment alternative, three most familiar steps should be considered are the required sewage quality, factors affecting management and applications namely: environmental, economic, and land availability (Tsagarakis et al. 2001; Sperling and Chernicharo 2005). The appropriateness of natural wastewater treatment alternatives could mainly be ranked by the effectiveness of system efficiency, minimum specific footprint to minimize land requirements, minimum energy utilization, costs (operation and maintenance), and little or no subjectivity to seasonal or climate changes (Qian et al. 2007; Srdjevic et al. 2017). Moreover, different studies compared various natural wastewater treatment methods through different models have mainly focused on the above mentioned or closely related criteria. For instance, Ouyang et al. (2015) evaluated slow rate land treatment, rapid infiltration land treatment, overland flow treatment, stabilization pond and constructed wetland using the above mention comparison criteria and Zeng et al. (2007) compared triple oxidation ditch, sequencing batch reactor and anaerobic single oxidation ditch by hierarchy gray relation method using land size, removal of phosphorous and nitrogen pollutants, plant stability and sludge disposing effects. In the current study, five natural wastewater treatment techniques such as waste stabilization pond (SP), constructed wetland (CW), use of aquatic plants (AP), soil filter (SF) and reuse of wastewater for irrigation (RWI) using land requirements (LR), capital cost (CC), pollutants removal efficiency (PRE), maintenance cost (MC), hydrogeological risks (HGR) and health risk as comparison criteria. The criteria in this study were mainly selected to focus on the environment, health, and economic issues. Economic factors are valuated based on maintenance, operation, and treatment costs, also land is evaluated by economic terms and costs in natural wastewater treatment methods is very cheap because most of the system components are operated without external source energy (Crites et al. 2006; Kivaisi 2001). Health-related issues are the basic health problems for the people living around wastewater treatment areas or peoples working onsite which originated from wastewater such as heavy metals, toxins, and other pathogens. The environmental factors focus on the potential impacts of the processes on hydrology, lithology, and ecology. Natural wastewater treatment methods are usually carbon-free and it is a smart-wastewater treatment method. For the wastewater treatment method to be classified as the best, it must be low cost, carbon emission-free/minimum carbon emission, and guarantee environmental safety and sustainability.

11.4 Results

In this study, VIKOR, a multi-criteria decision-making technique method has been applied for the analysis of the natural wastewater treatment methods performance. For this aim firstly the decision matrix of the natural wastewater treatment methods has been constructed and obtained (Table 11.2).

The importance weights of each parameter for fundamental scales of comparison selected by the experts are given in Table 11.3 in the linguistic scale and triangular fuzzy scale.

And the importance weights of the parameters defuzzified using the Yager index to obtain a single value out of the triangular fuzzy numbers and normalized for the use in the VIKOR technique. Then the best (f_i^*) and the worst values (f_i^-) sets of the alternatives based on each criterion have been calculated as shown in Table 11.4.

The utility (S_j) and the regret (R_j) measures of each alternative have been computed for each natural wastewater treatment methods (see Table 11.5). The natural wastewater treatment methods were compared based on comparison criteria and weight selected. The natural wastewater treatment methods addressed in this study have indicated different performance levels when evaluated with similar comparison criteria (see Table 11.6). The criteria are selected predominantly based on the environmental, economic, and health impacts of the wastewater treatment methods. The ranking of the treatment methods was computed for the cumulative effects of all evaluation criteria. The results indicated (see Table 11.6) that the stabilization pond is the first ranked and the reuse of wastewater for irrigation is last ranked. Hence, it is logical to conclude waste stabilization pond is the best natural wastewater treatment method among the methods addressed in this study. The finding of this study is in agreement with (Gichamo et al. 2020; Ouyang et al. 2015). Furthermore, the performance of waste stabilization pond can diminish BOD_5, suspended solid ammonia nitrogen, and phosphorous up to 90%, 93% 91%, and 84% respectively, (Surampalli et al. 2007). The other study by waste stabilization pond ranked at first thus, achieved 94% removal of total nitrogen, 93% of nitrate nitrogen, and 96% of ammonia nitrogen (Duan and Fedler 2010).

In VIKOR modeling the alternative with minimum Q_i value is the best alternative and the one with maximum Q_i is the worst alternative. The result (Table 11.6) indicated according to the order of best to worst alternatives measured by the balanced effects of a given criterion. For all the given evaluation criteria, the waste stabilization pond was observed as an appropriate wastewater treatment method because the method meets the objective of each evaluation criterion. The method is independent of seasonal variability and related climatic variations, the overall cost of the method is fair and it did not cause/cause a little negative health impact on humans working on-site or living around as compared with other methods.

The utility (S_j) and the regret (R_j) measures of each alternative have been calculated and presented (Table 11.5).

Table 11.2 Decision matrix of the natural wastewater treatment methods

Alternatives/criteria	Pollutants removal efficiency	Land requirement	Capital cost	Maintenance cost	Health risk	Hydrogeological risk	Ecological benefit	Subjected to seasonal effects
Aim (max/min)	Max	Min	Min	Min	Min	Min	Max	Min
Importance of weight	VH	VH	VH	VH	VH	VH	VH	H
Stabilization pond	9	5	4	3	2	3	8	3
Constructed wetland	8	6	6	4	3	3	8	6
Use of aquatic plants	9	6	6	6	4	2	7	7
Soil filters	5	6	4	2	3	5	4	4
Reuse of wastewater for irrigation	6	9	5	4	7	7	4	8

Table 11.3 Linguistic fuzzy Scale

Linguistic scale for importance of weight	Triangular fuzzy scale
Very high (VH)	(0.75, 1, 1)
Important (H)	(0.50, 0.75, 1)
Medium (M)	(0.25, 0.50, 0.75)
Low (L)	(0, 0.25, 0.50)
Very low (VL)	(0, 0, 0.25)

The ranking result of VIKOR analysis obtained based on the Q_i index, which is a combination of S_i and R_i indexes. The alternative with a lower Q_i is the better alternative and the alternative with higher Q_i is the least desirable alternative (see Table 11.6). Results indicated that the stabilization pond technique is the best alternative amongst the others and Constructed wetland is the second-best alternative.

11.5 Conclusions

Advancement of industry and population increasing rise demands for raw materials and subsequent utilization increases the generation of wastewater. Unsafe disposal of wastewater causes pollution of the environment disturbs human health and economic crisis. To reduce the negative consequences of wastewater, it could be treated and the pollutants must be removed at an acceptable limit before disposing to the environment. Wastewater could be treated either by natural treatment methods or conventional treatment methods; the latter is economically more expensive and technically needs advanced skills. Natural wastewater treatment techniques are cheaper, free from carbon emission, could be constructed, operated, and maintained by little skill. The current study focused on assessing the effectiveness of constructed wetlands, soil filter, waste stabilization basin, reuse of wastewater for irrigation and aquatic plants using VIKOR model. The result reveals that waste stabilization pond was the first ranked and an appropriate method, constructed wetland, soil filter, use of aquatic plants and reuse of wastewater for irrigation is the second-ranked the second, the third, the forth and the fifth respectively. In VIKOR, the best alternative is the one with the minimum value of Q_i. The Q_i value for waste stabilization is the least; therefore, this method of natural wastewater treatment is the best option among the alternatives evaluated. This could be because the method balances the cost, environmental and human health issues at a desirable level.

Table 11.4 The positive (f_i^*) and the negative ideal points (f_i^-) of the alternatives based on each criterion

Alternatives/criteria	Pollutants removal efficiency	Land requirement	Capital cost	Maintenance cost	Health risk	Hydrogeological risk	Ecological benefit	Subjected to seasonal effects
f_i^*	9	5	4	2	2	2	8	8
f_i^-	5	9	6	6	7	7	4	3

Table 11.5 S_j and R_j indexes of the natural wastewater treatment methods

Alternatives	S_i	R_i
Stabilization pond	0.0576	0.0320
Constructed wetland	0.3697	0.1280
Use of aquatic plants	0.4545	0.1280
Soil filters	0.4111	0.1280
Reuse of wastewater for irrigation	0.8401	0.1280

Table 11.6 Ranking results of the natural wastewater treatment methods

Rank	Alternatives	Q_i
1	Stabilization pond	0.0000
2	Constructed wetland	0.6994
3	Soil filters	0.7259
4	Use of aquatic plants	0.7536
5	Reuse of wastewater for irrigation	1.0000

References

Akber A, Mukhopadhyay A, Al-Senafy M, Al-Haddad A, Al-Awadi E, Al-Qallaf H (2008) Feasibility of long-term irrigation as a treatment method for municipal wastewater using natural soil in Kuwait. Agric Water Manag 95(3):233–242. https://doi.org/10.1016/j.agwat.2007.09.015

Bottero M, Comino E, Riggio V (2011) Application of the Analytic Hierarchy Process and the Analytic Network Process for the assessment of different wastewater treatment systems. Environ Model Softw 26(10):1211–1224. https://doi.org/10.1016/j.envsoft.2011.04.002

Corcoran E, Nelleman C, Baker E, Bos R, Osborn D, Savelli H (2010) Sick water? The central role of wastewater management in sustainable development. Water. https://doi.org/10.1007/s10230-011-0140-x

Crites RW, Middlebrooks EJ, Bastian RK (2014b) Natural wastewater treatment systems (2nd edn) https://doi.org/10.1201/b16637

Crites RW, Middlebrooks EJ, Reed SC (2006) Natural wastewater treatment systems. In: Natural wastewater treatment systems. https://doi.org/10.1201/b16637-2

Duan R, Fedler CB (2010) Performance of a combined natural wastewater treatment system in west texas. J Irrig Drain Eng 136(3):204–209. https://doi.org/10.1061/(ASCE)IR.1943-4774.0000154

Dubey AK, Sahu O (2014) Review on natural methods for waste water treatment. J Urban Environ Eng 8(1):89–97. https://doi.org/10.4090/juee.2014.v8n1.089097

Duckstein L, Opricovic S (1980) Multiobjective optimization in river basin development. Water Resour Res 16(1):14–20

Gersberg RM, Elkins BV, Lyon SR, Goldman CR (1986) Role of aquatic plants in wastewater treatment by artificial wetlands. Water Res 20(3):363–368. https://doi.org/10.1016/0043-1354(86)90085-0

Gichamo T, Gökçekuş H (2019) Interrelation between climate change and solid waste. J Environ Pollut Contr 2(1):1–7

Gichamo T, Gökçekuş H, Ozsahin DU, Gelete G, Uzun B (2020) Evaluation of different natural wastewater treatment alternatives by fuzzy PROMETHEE method. Desalin Water Treatm 177(May 2019):400–407. https://doi.org/10.5004/dwt.2020.25049

Ho L, Echelpoel W, Van Charalambous P, Gordillo APL, Thas O, Goethals P (2018) Statistically-based comparison of the removal efficiencies and resilience capacities between conventional and natural wastewater treatment systems: a peak load scenario. Water (Switzerland), 10(3). https://doi.org/10.3390/w10030328

Hunter RG, Day JW, Wiegman AR, Lane RR (2018) Municipal wastewater treatment costs with an emphasis on assimilation wetlands in the Louisiana coastal zone. Ecol Eng (January), 0–1. https://doi.org/10.1016/j.ecoleng.2018.09.020

Kim Y, Chung ES, Jun SM, Kim SU (2013) Prioritizing the best sites for treated wastewater instream use in an urban watershed using fuzzy TOPSIS. Resour Conserv Recycl 73(1):23–32

Kivaisi AK (2001) The potentail for constructed wetlands for wastewater treatmen and resuse in developing countries: a review. Ecol Eng 16:545–600

Kuschk P, Müller RA, Moormann H, Kästner M, Bederski O, Wießner A, Stottmeister U (2003) Effects of plants and microorganisms in constructed wetlands for wastewater treatment. Biotechnol Adv 22(1–2):93–117. https://doi.org/10.1016/j.biotechadv.2003.08.010

Laitinen J, Moliis K, Surakka M (2017) Resource efficient wastewater treatment in a developing area—climate change impacts and economic feasibility. Ecol Eng 103:217–225. https://doi.org/10.1016/j.ecoleng.2017.04.017

Mitsch WJ (2012) What is ecological engineering? Ecol Eng 45(October):5–12. https://doi.org/10.1016/j.ecoleng.2012.04.013

Ouyang X, Guo F, Shan D, Yu H, Wang J (2015) Development of the integrated fuzzy analytical hierarchy process with multidimensional scaling in selection of natural wastewater treatment alternatives. Ecol Eng 74:438–447. https://doi.org/10.1016/j.ecoleng.2014.11.006

Qian Y, Wen X, Huang X (2007) Development and application of some renovated technologies for municipal wastewater treatment in China. Front Environ Sci Eng China 1(1):1–12. https://doi.org/10.1007/s11783-007-0001-9

Rozkosny M, Kriska M, Salek J, Bodik I, Istenic D (2014) Natural technologies of wastewater treatment

Sadr SMK, Saroj DP, Mierzwa JC, McGrane SJ, Skouteris G, Farmani R, Ouki SK (2018) A multi expert decision support tool for the evaluation of advanced wastewater treatment trains: a novel approach to improve urban sustainability. Environ Sci Pol 90(September):1–10. https://doi.org/10.1016/j.envsci.2018.09.006

Srdjevic B, Srdjevic Z, Suvocarev K (2017) Multi-criteria evaluation of wastewater treatment technologies in constructed wetlands. Eur Water 58(3):165–171

Sun Y, Garrido-Baserba M, Molinos-Senante M, Donikian NA, Poch M, Rosso D (2020) A composite indicator approach to assess the sustainability and resilience of wastewater management alternatives. Sci Total Environ 725:138286. https://doi.org/10.1016/j.scitotenv.2020.138286

Surampalli RY, Banerji SK, Tyagi RD, Yang PY (2007) Integrated advanced natural wastewater treatment system for small communities. Water Sci Tech 55(11):239–243. https://doi.org/10.2166/wst.2007.371

Tsagarakis KP, Mara DD, Angelakis AN (2001) Wastewater management in Greece: experience and lessons for developing countries. Water Sci Technol 44(6):163–172

Tsaur SH, Chang TY, Yen CH (2002) The evaluation of airline service quality by fuzzy MCDM. Tourism Manage 23:107–115

Von Sperling M, Chernicharo CADL (2005) Biological wastewater treatment in warm climate regions. IWA Publishing, pp 1–856. https://doi.org/10.5860/CHOICE.45-2633

Yu PL (1973) A class of solutions for group decision problems. Manage Sci 19(8):936–946

Zeng G, Jiang R, Huang G, Xu M, Li J (2007) Optimization of wastewater treatment alternative selection by hierarchy grey relational analysis. J Environ Manage 82(2):250–259. https://doi.org/10.1016/j.jenvman.2005.12.024

Chapter 12
Evaluating Disinfection Techniques of Water Treatment Using Multi-criteria Decision-Making Method

Gebre Gelete, Hüseyin Gökçekuş, Berna Uzun, Dilber Uzun Ozsahin, and Tagesse Gichamo

Abstract Disinfection in the process of water treatment is used to make water safe for intended purpose by eliminating pathogenic microorganisms from it. There are different kinds of disinfection methods used in water supply each of them with their own weakness and strength. The main objective of this study is to evaluate different water disinfection methods using VIKOR model and selecting the best alternative among the computing disinfection techniques. The study compares five water disinfection methods namely Ozone, Ultraviolent radiation, chlorination, chlorine dioxide and chloramination. For comparison of the disinfection methods, different

G. Gelete (✉) · H. Gökçekuş · T. Gichamo
Faculty of Civil and Environmental Engineering, Near East University, Nicosia, Turkish Republic of Northern Cyprus, Turkey
e-mail: gegelete04@gmail.com

H. Gökçekuş
e-mail: huseyin.gokcekus@neu.edu.tr

T. Gichamo
e-mail: tagesseg@gmail.com

G. Gelete · T. Gichamo
College of Environmental Science, Arsi University, 193 Asela, Ethiopia

B. Uzun · D. Uzun Ozsahin
DESAM Institute, Near East University, Nicosia, Turkish Republic of Northern Cyprus, Turkey
e-mail: berna.uzun@neu.edu.tr

D. Uzun Ozsahin
e-mail: dilber.uzunozsahin@neu.edu.tr

B. Uzun
Department of Mathematics, Near East University, Nicosia, Turkish Republic of Northern Cyprus, Turkey

D. Uzun Ozsahin
Department of Biomedical Engineering, Near East University, Nicosia, Turkish Republic of Northern Cyprus, Turkey

Medical Diagnostic Imaging Department, College of Health Science, University of Sharjah, Sharjah, United Arab Emirates

101

evaluation criteria such as capital cost, maintenance and operation cost, pathogen removal efficiency, system reliability, operational simplicity, safety risk, disinfectant residue formation and undesirable byproduct formation was used. The VIKOR analysis showed that ultraviolent radiation is the most preferable disinfection method and chlorine dioxide is the least preferable disinfectant.

Keywords Disinfection · VIKOR

12.1 Introduction

Water is one of vital and abundant resource on the earth surface which all living beings are depends ant on it. However, the occurs contamination of water resources because of improper waste water disposal to water bodies, industrialization and increasing population (Gelete et al. 2020). Fresh water is depleting and limited. Getting access for safe drinking water is the main issue worldwide particularly in rural areas and developing countries. Millions of people globally are exposed to waterborne disease because of lack of access to safe drinking water (Song et al. 2016). Therefore, water has to be treated before supplying it to the consumers. During water treatment, water must pass through a number of steps to get rid of from chemical, physical and biological pollutants (Gelete et al. 2020).

Disinfection is one of those stages of water treatment which protects human health by inactivates and kills waterborne and harmful microorganisms in the drinking water before it is being supplied for the consumers (Huertas et al. 2008). Thus, it is vital step in treatment process of municipal water supply due to its importance in removing biological impurities from drinking water (Sun et al. 2019). Disinfection is commonly considered as the standard treatment process of drinking water, swimming pool and wastewater treatment. The commonly used disinfection techniques used in drinking water includes ultraviolet radiation, chlorine, ozonation, chloramine and chlorine dioxide (Richardson 1998). These methods are widely used throughout the world even though each of the have their own weakness and strengths. Also, water users have concerns about the reliability, efficiency and toxic byproduct formed from disinfection process (Gelete et al. 2020). Therefore, designing of water treatment system and selecting the appropriate disinfection method is one of the important task of engineers (Avramenko et al. 2010).

Selecting the best method of drinking water disinfection before designing and implementing of any water treatment plant is crucial. If there are different water disinfection methods, it is challenging for the decision makers to decide which one is to be selected because of multitude of criteria's which is a multi-criteria decision making problem. There are different methods used for ranking and selection of the best among the alternatives. Multi-criteria decision-making methods such VlseCriterion Optimization and Compromise Solution in Serbian (VIKOR) can be successfully used to select the best alternative among the competing methods. The VIKOR technique was devoted foe multi-criteria complex problems optimization.

Anojkumar et al. (2014) compared the ranking performances of different multi-criteria decision making (MCDM) techniques (viz. VIKOR, TOPSIS, PROMETHEE and ELECTRE) for selection of pipe materials and found that the application of VIKOR is valuable assistant in decision making problems.

Therefore, the main objective of the study is to analyze different techniques of water disinfection using Multi-criteria decision making technique by VIKOR to select the best alternative among the water disinfection methods. There are different options to achieve the objective of water disinfection process. Therefore, these alternative should be compared and evaluated against different criteria's such as economic, social and effectiveness using different techniques. The process of selecting the alternatives should include cost, effectiveness of the method and service life of the system.

12.2   Water Disinfection Techniques

12.2.1   Ultraviolet Radiation (UVR)

The first application of UVR irradiation in drinking water as disinfection process was started in 1910 in Marseille (Hijnen et al. 2006). Nowadays, its usage is getting a growing interest for tertiary disinfection in drinking water treatment. In UVR disinfection process, the water is exposed to short wave radiation so as to kill and inactivate the disease causing microorganisms (EPA 2011).

At present, ultraviolet radiation is a widely used disinfectant in water treatment due to its capacity to inactivate a variety of disease-causing microorganisms (Chatzisymeon et al. 2011). The use of UVR for water disinfection has been increasing as it is effective method to kill and in activate different microorganisms in water. It has several advantages over chemical disinfection methods (e.g., Ozonation, Chlorination), such as no formation of harmful disinfection byproduct, no addition of chemical, less cost, straightforward to maintain, and is an incredibly rapid process (Mori et al. 2007). Most researchers have been recommended UVR disinfection to substitute for other chemical disinfection methods in water treatment (Brownell et al. 2008). Despite what might be expected, the lack of residual disinfection is the main disadvantage of utilizing UVR. Disinfection innovation is supposed to be perfect when it is financially savvy and doesn't create side effects (e.g., harmful byproduct) (Chen et al. 2007).

Ultraviolet radiation is a compelling disinfectant, and it doesn't impact the quality of water being treated. This is the fact that ultraviolet radiation is a physical method for removing microorganisms, i.e., no chemical is used for disinfecting water, and the water doesn't experience any chemical change. Subsequently, the pH, taste, and smell are not changed, as the main objective is only the pathogens (Gelete et al. 2020). Notwithstanding drinking water treatment, this strategy can likewise be applied in the purification of treated wastewater.

12.2.2   Chlorination (CL)

Removing disease causing microorganisms from water is the main objective of disinfection in water treatment process. Chlorination is the most commonly used method worldwide to achieve this objective in municipal water supply. In the chlorination process chlorine or chlorine byproducts are added to the water and reacts with water to form hypochlorite ions and hypochlorous acid (Pichel et al. 2019).

Chlorination is effective and cheap disinfection technique even at low concentration and no post treatment is required as it forms a residual in water distribution system. The primary advantage of chlorination is that the residual chlorine lasts longer in the distribution system. Therefore, there is no possibility of recontamination of water during distribution and storage (EPA 2011). Chlorination has been widely used around world as water disinfection method due to its effectiveness of pathogen removal and low cost. However, chlorination has many disadvantages, such as the undesirable taste and odor, formation of trihalomethanes byproducts, inadequacy against protozoa eggs and cysts, the, and in excess of 400 different sorts of chlorination byproducts (Zhai et al. 2017). The other issue related with chlorination is that there is no fixed principle on the amount of chlorine required to be added. In any case, the mount required relies upon the water quality and the disinfection necessity. Besides, a water treatment plant that utilizes chlorine gas as the disinfectant requires exceptionally skilled operators, maintenances and man power (Gadgil 1998). However, a treatment plant that uses diluted chlorine is relatively cost-effective and straightforward. Nevertheless, the worldwide applicability of this method can be ascribed to its convenience and to its exceedingly acceptable performance as a disinfectant, which has been built up by many years of usage.

12.2.3   Chloramination (CM)

In the chloramination process, ammonia and chlorine are mixed and react to form monochloramine. This process should be done under well controlled conditions to prevent strong tastes and byproducts formation (Chatzisymeon et al.2011; EPA 2011). The efficiency of monocloramin in removing microorganism in water is low compared to chlorination and hence it is mostly used to deliver a residual disinfectant during distribution and storage of treated water.

Utilizing chloramination for drinking water disinfection is advantageous because of no harmful byproducts (e.g. trihalomethanes) is formed under the present of organic matter. Additionally, the taste limit is commonly a lot higher than of chlorine alone. Accordingly, utilizing chloramination in disinfecting drinking water can altogether diminish client protests related with chlorine tastes. Because of this explanation, the utilization of chloramination water disinfection is getting progressively well known in most countries as it gives residual chlorine in distribution lines. The residual chlorine in the distribution system prevents the water from recontamination. On the other hand, chloramination has some drawbacks. This includes requires chemical, requires skilled manpower, less efficient in removing microorganisms.

12.2.4 Ozonation (OZ)

Ozone is an excellent disinfectant and oxidant that is widely used in municipal water supply attain disinfection improvement and water quality (Gunten 2003). Ozone is produced onsite by passing dry air through a system of high voltage electrodes (Summerfelt 2003). Ozone is an effective disinfectant in water treatment and can even inactivate very resistant microorganisms (e.g. *Cryptosporidium* and *Giardia.*) which conventional methods fail.

Ozone is an excellent disinfectant and can even be used to inactivate many kinds of microorganisms such as protozoa which are very resistant to conventional disinfectants. Ozone is more effective than chlorination and chlorine dioxide as it readily oxidizes pesticides, chemical residuals, organic matter and wide ranges of microorganisms at low concentration and in short time (EPA 2011; Pichel et al. 2019). Moreover, this method is primarily effective in removing spores and cysts.

The limitation of ozone for drinking water disinfection is that there could be recontamination of water during distribution line and storage as its concentration decays rapidly. Moreover, using ozone for water disinfection is expensive especially in terms of capital and operational costs. It requires highly skilled manpower for onsite generation, for maintenance, high energy requirement, and post-treatment to remove organic carbon formed during the oxidation process (EPA 2011). Ozone exposure to inactivate pathogen is quite high which leads undesirable disinfection byproduct formation particularly bromate (Gunten 2003).

12.2.5 Chlorine Dioxide (CD)

Chlorine dioxide (ClO_2) is one of the method used for disinfection of drinking water specially to remove algae from the water. It can also effectively remove taste, odor, manganese and taste from drinking water. Chlorine dioxide is unstable and thus it is sensitive to light, temperature and pressure. Chlorine dioxide is highly explosive in the air especially when the concentration is above 4%.

12.3 Materials and Methods

In this study, different water disinfection techniques are analyzed and ranked using VIKOR Multi-criteria decision making techniques considering different evaluation criteria's.

12.3.1 Evaluation Criteria

The selection of the best water disinfection method is critical and is always conducted by considering different criteria. Different disinfection methods are therefore compared by setting different criteria. In this study the criteria considered mainly includes technical performance, economic feasibility and suitability of the technique. Different indices can be used by the decision makers to compare various items (Von Sperling and Oliveira 2010). In this work, reliability, capital cost, operational and maintenance cost, operational simplicity, pathogen removal efficiency, undesirable byproduct formation and safety risk are used as criteria for evaluation of water disinfection methods.

12.3.2 A Capital Cost (CC) and Maintenance and Operation Cost (MOC)

Economic criteria (e.g. cost) is important indices in evaluating different alternatives. The types of costs in this study considered are capital, operational and maintenance costs. CC are those costs related to cost of land, installation and construction of electromechanical works and equipment's which required for implementing the whole system (Karimi et al. 2010; Pophali et al. 2011). Whereas, MOC consists costs related to maintain and operation process such as repair costs, energy cost personnel costs and chemical product costs (Avramenko et al. 2010). MOC and CC have been widely used in many studies for ranking of wastewater treatment methods (Curiel-Esparza et al. 2014; Karimi et al. 2010; Zeng et al. 2007) and water disinfection (Gelete et al. 2020).

12.3.3 Reliability of the System (RS)

Reliability of the disinfection system is a crucial criterion for evaluating different kinds of water disinfection methods (Crites and Tchobanoglous 2000). In this study, reliability refers to the ability of disinfection method for achieving satisfactory performances for a certain condition for specific time period. Reliability in water disinfection technique can be viewed as the percentage to which treated water quality complies with World Health organization (WHO) standard (Niku et al. 1982). To be selected as the reliable system, water disinfection technique must give highest performance and effectiveness during normal and emergencies operations. Hence, in water treatment reliability of disinfection methods are measured with regard to effluent quality and consistency with drinking water standards (Karimi et al. 2010).

12.3.4 Operational Simplicity (OS)

Comparison of disinfection method should consider the operational and maintenance need of the system. If disinfection process requires skilled workers especially in remote areas, decision makers can reject it. Curiel-Esparza et al. (2014) As the long-term success of the system is determined by simplicity (Karimi et al. 2010), maintenance and operational simplicity should be the main issue in ranking and selecting the disinfection methods.

12.3.5 Pathogen Removal Efficiency (PRE)

The water treated using different treatment process should be disinfected until the final effluent comply with the WHO standard for drinking water (EPA 2011). This criterion evaluates the capability of disinfection method to eliminate pathogens and contaminants from the water. The pathogen removal efficiency of water disinfection process can be assessed from full treatment plant and its past experience studies (Karimi et al. 2010). So far many study has been used pathogen removal efficiency criteria for evaluating of water and wastewater disinfection alternatives (Gelete et al. 2020; Grunert et al. 2018).

12.3.6 Undesirable Byproduct Formation (UBF)

When disinfection chemicals react with other organic or inorganic chemicals in water disinfection products will be formed. According to these byproducts have adverse human health effect if consumed (Lantagne et al. 2010; Smith et al. 2010). Therefore, UBF is used in this study as one of the main criteria's t evaluate and rank different water disinfectant techniques.

12.3.7 Safety Risk (SR)

Workers safety during water treatment process should be taken into consideration in evaluation and selection of the best among computing disinfection methods. This is due to the fact that bad accident could happen on the workers during the operation and maintenance of the system.

12.3.8 Residue Formation

RF in water treatment is the ability of water disinfection method to create residues that protects the treated water from recontamination because of short coming in the water storage and distribution system (Curiel-Esparza et al. 2014). Thus, evaluation should be made to make sure that disinfectant residue is available in the water storage and distribution system (EPA 2011). The best disinfectant alternative is the one which persists for long even after treatment.

12.4 Application of the Fuzzy Based-VIKOR Technique for the Analysis of the Water Disinfection Methods

In this study, fuzzy based-VIKOR which is a known MCDM method is used to analyze the performance of different water disinfection methods.

For evaluation of disinfectant methods considering different criteria, firstly the decision matrix of the disinfection methods has been obtained as seen in Table 12.1.

The importance weights of each parameter selected by the experts with the linguistic fuzzy scale as presented in Table 12.2.

Table 12.1 Decision matrix of the disinfection methods

Alternatives/Criteria	CC	OMC	OS	PRE	SR	RS	RF	UBF
Aim (max/min)	Min	Min	Max	Max	Min	Max	Max	Min
Importance weight	VH	VH	VH	VH	VH	VH	H	VH
UVR	5	3	9	9	4	9	4	1
CL	6	5	6	7	8	7	7	8
CM	7	6	4	2	5	5	7	6
OZ	9	8	7	8	6	8	4	2
CD	7	7	5	5	8	6	6	6

Table 12.2 Linguistic Fuzzy Scale

Linguistic scale for importance weight	Triangular fuzzy scale
Very high (VH)	(0.75, 1, 1)
Important (H)	(0.50, 0.75, 1)
Medium (M)	(0.25, 0.50, 0.75)
Low (L)	(0, 0.25, 0.50)
Very low (VL)	(0, 0, 0.25)

12.5 Results and Discussion

The application of VIKOR model in evaluation and ranking of different water disinfection technique gives accurate result due to the fact that consistent ranking output is obtained. And the importance weights of the parameters defuzzified using the Yager index to obtain a single value out of the triangular fuzzy numbers and normalized for the use of the VIKOR technique. Then the best (f_i^*) and the worst values (f_i^-) sets of the alternatives based on each criterion have been calculated as shown in Table 12.3.

And the utility (S_j) and the regret (R_j) measures of each alternative have been calculated as shown in Table 12.4.

Lastly the ranking result of this analysis obtained based on the (Q_i) index, which is a combination of the S_i and R_i indexes. The alternative with a lower (Q_i) is the better alternative and the ranking result is presented in Table 12.5.

Table 12.5 depicts the ranking result of different water disinfection methods based on the considered criteria's. Results show that, the UVR is the best alternative amongst the others with a valuable differences and CD is the least alternative. Table 12.5 depicts the evaluation and ranking of different water disinfection techniques depending on the applied weight and criteria. The analysis result revealed that UVR is the best and CD is the least preferable disinfection methods. This is because UVR requires minimum MOC, has higher reliability, highest PRE, no unwanted

Table 12.3 The positive (f_i^*) and the negative ideal points (f_i^-) of the alternatives

Alternatives/criteria	CC	OMC	OS	PRE	SR	RS	RF	Error! not a valid link
f_i^*	5	3	9	9	4	9	7	1
f_i^-	9	8	4	2	8	5	4	8

Table 12.4 S_j and R_j indexes of the disinfection methods

Alternatives	S_i	R_i
UVR	0.75	0.75
CR	3.712857143	0.92
CM	4.659142857	0.92
OZ	3.910857143	0.92
CD	4.974857143	0.92

Table 12.5 Ranking result of the disinfection methods

Rank	Alternatives	Q_i
1	UVR	0
2	CR	0.850645838
3	CM	0.874078583
4	OZ	0.962636099
5	CD	1

byproduct formation, and simple to operate. In the contrary because of its high safety risk, lower operational simplicity, low efficiency in removing pathogen and unwanted byproduct formation CD is ranked last among the competing alternatives.

12.6 Conclusion

This study used VIKOR model to evaluate and select the best drinking water disinfection methods by considering different evaluation criterias in the evaluation of disinfection performances, the economic and technical criteria were considered. The analysis result showed that UVR is the best alternative among the competing dis infection methods an CD is the least. Therefore, it can be concluded that choosing ultraviolet radiation technology seems to give the best overall account of technical and economic concerns.

References

Anojkumar L, Ilangkumaran M, Sasirekha V (2014) Comparative analysis of MCDM methods for pipe material selection in sugar industry. Expert Syst Appl 41(6):2964–2980. https://doi.org/10.1016/j.eswa.2013.10.028

Avramenko Y, Kamami M, Kraslawskia A (2010) Fuzzy performance indicators for decision making in selection of wastewater treatment methods. Comput Aided Chem Eng 28(127):132

Brownell SA, Chakrabarti AR, Kaser FM, Connelly LG, Peletz RL, Fermin R, Lang MJ, Kammen DM, Nelson KL (2008) Assessment of a low-cost, point-of-use, ultraviolet water disinfection technology. J Water Health 6(1):53–65

Chatzisymeon E, Droumpali A, Mantzavinos D, Venieri D (2011) Disinfection of water and wastewater by UV-A and UV-c irradiation: application of real-time PCR method. Photoch Photobio sci 10(3):389–395

Chen JP, Yang L, Wang LK, Zhang B (2007) Ultraviolet Radiation for Disinfection. Adv Physicochem Treatm Process 4:317–366

Crites R, Tchobanoglous G (2000) Tratamiento de Aguas Residuales En Pequeñas Poblaciones. McGraw-Hill Interamericana S.A., Bogotà, Colombia, p 776

Curiel-Esparza J, Cuenca-Ruiz MA, Martin-Utrillas M, Canto-Perello J (2014) Selecting a sustainable disinfection technique for wastewater reuse projects. Water (Switzerland) 6(9):2732–2747

Huertas E, Salgot E, Hollender J, Weber S, Dott W, Khan S, Schäfer A, Messalem R, Bis B, Aharoni A, Chikurel H (2008) Key objectives for water reuse concepts. Desalination 218(1–3):120–131

EPA (2011) Water treatment manual: disinfection. ISBN: 978-184095-421-0

Gadgil A (1998) Drinking water in developing countries. Ann Rev Energy Environ 23:253–286

Gelete G, Gokcekus H, Uzun D, Uzun B, Gichamo T (2020) Evaluating disinfection techniques of water treatment. Desal Water Treat 177(May 2019): 408–15

Grunert A, Frohnert A, Selinka H, Szewzyk R (2018) A New approach to testing the efficacy of drinking water disinfectants. Int J Hyg Environ Health 221(8):1124–1132. https://doi.org/10.1016/j.ijheh.2018.07.010

Gunten UV (2003) Ozonation of drinking water: part II. disinfection and by-product formation in presence of bromide, iodide or chlorine. Water Research 37:1469–1487

Hijnen WAM, Beerendonk EF, Medema GJ (2006) Inactivation credit of uv radiation for viruses, bacteria and protozoan (Oo)cysts in water: a review. Water Res 40(1):3–22

Karimi A, Mehrdadi N, Hashemian S, Bidhendi G, Moghaddam R (2010) Selection of wastewater treatment process based on the analytical hierarchy process and fuzzy analytical hierarchy process methods. Int J Environ Sci Technol 8:267–280

Karimi AR, Mehrdadi N, Hashemian SJ, Nabi Bidhendi GR, Tavakkoli Moghaddam R (2011) Selection of wastewater treatment process based on analytical hierarchy process. Int J Civil Eng Tech 5(9):27–33

Lantagne DS, Cardinali F, Blount B (2010) Disinfection by-product formation and mitigation strategies in point-of-use chlorination with sodium dichloroisocyanurate in tanzania. Am J Trop Med Hyg 83(1):135–143

Mori M, Hamamoto A, Takahashi A, Nakano M, Wakikawa N, Tachibana S, Ikehara T, Nakaya Y, Akutagawa M, Kinouchi Y (2007) Development of a new water sterilization device with a 365 Nm UV-LED. Biol. Eng. Comput. 45(12):1237–1241

Niku S, Schroeder ED, Haugh RS (1982) Reliability and stability of trickling filter processes. J Water Pollut Contr Assoc 54(2):129–34

Pichel N, Vivar M, Fuentes M (2019) The problem of drinking water access: a review of disinfection technologies with an emphasis on solar treatment methods. Chemosphere 218:1014–1030. https://doi.org/10.1016/j.chemosphere.2018.11.205

Pophali GR, Chelani AB, Dhodapkar RS (2011) Optimal selection of full scale tannery effluent treatment alternative using integrated AHP and GRA approach. Expert Syst Appl 38(9):10889–10895. https://doi.org/10.1016/j.eswa.2011.02.129

Richardson S (1998) Drinking water disinfection by-products. Encyclop environ Anal. Reme 3:1398–1421

Smith EM, Plewa MJ, Lindell CL, Richardson SD, Mitch WA (2010) Comparison of byproduct formation in waters treated with chlorine and iodine: relevance to point-of-use treatment. Environ Sci Technol 44(22):8446–8452

Song K, Mohseni M, Taghipour F (2016) Application of ultraviolet light-emitting diodes (UV-LEDs) for water disinfection: a review. Water Res 94:341–349. https://doi.org/10.1016/j.watres.2016.03.003

Von Sperling M, Oliveira (2010) Reliability analysis of stabilization ponds systems. Water Sci Technol 55:127–134

Summerfelt ST (2003) Ozonation and UV irradiation—an introduction and examples of current applications. Aquacult Eng 28(1–2):21–36

Sun, X, Chen M, Wei D, Du Y (2019) Research progress of disinfection and disinfection by-products in China. Int J Environ Sci https://linkinghub.elsevier.com/retrieve/pii/S100107421832850X

Zeng G, Jiang R, Huang G, Xu M, Li J (2007) Optimization of wastewater treatment alternative selection by hierarchy grey relational analysis. J Environ Manage 82(2):250–259

Zhai H, He X, Zhang Y, Du T, Adeleye AS, Li Yao (2017) Chemosphere disinfection byproduct formation in drinking water sources: a case study of Yuqiao reservoir. Chemosphere 181:224–231. https://doi.org/10.1016/j.chemosphere.2017.04.028

Chapter 13
Evaluation of the Learning Models Using Multi-criteria Decision Making Theory

Gülsüm Aşıksoy, Berna Uzun, and Dilber Uzun Ozsahin

Abstract Many different learning methods are used in education, and some are more effective than others the aim of this study is to evaluate the performance of various types of learning methods by applying multi-criteria decision making analysis. In this study, we analyzed the most commonly used learning methods such as traditional learning, flipped classroom method, collaborative learning, and the gamification approach in order to evaluate the performance of these methods based on important parameters. The study was conducted with 180 first year engineering students assigned to four groups. Group 1 learned with the traditional learning method, group 2 learned with the flipped classroom method, group 3 learned with collaborative learning, and group 4 learned with the gamified flipped classroom. Data were collected from a physics achievement test, science motivational questionnaire, physics attitude questionnaire, and class attendance list. The Fuzzy PROMETHEE method (Preference Ranking Organization Method for Enrichment of Evaluations) has been used, which is a multi-criteria decision-making technique successfully

G. Aşıksoy
Computer Education and Instructional Technology, Near East University, Nicosia, Turkish Republic of Northern Cyprus, Turkey

B. Uzun (✉) · D. Uzun Ozsahin
DESAM Institute, Near East University, Nicosia, Turkish Republic of Northern Cyprus, Turkey
e-mail: berna.uzun@neu.edu.tr

D. Uzun Ozsahin
e-mail: dilber.uzunozsahin@neu.edu.tr

B. Uzun
Department of Mathematics, Near East University, Nicosia, Turkish Republic of Northern Cyprus, Turkey

D. Uzun Ozsahin
Department of Biomedical Engineering, Near East University, Nicosia, Turkish Republic of Northern Cyprus, Turkey

Medical Diagnostic Imaging Department, College of Health Science, University of Sharjah, Sharjah, United Arab Emirates

© The Author(s), under exclusive license to Springer Nature Switzerland AG 2021
D. Uzun Ozsahin et al. (eds.), *Application of Multi-Criteria Decision Analysis in Environmental and Civil Engineering*, Professional Practice in Earth Sciences,
https://doi.org/10.1007/978-3-030-64765-0_13

applied in many fields. The evaluation results showed that the gamified flipped classroom method was the best option, while the traditional method was the least effective. This study provides valuable information about the learning methods.

Keywords Learning methods · Fuzzy PROMETHEE · Decision making

13.1 Introduction

Education is the most important force that has the power to change the world. The most effective way of achieving this is to educate individuals who can think, perform research, question and solve problems. Educational systems that produce conditioned, behavioral, imitative, and memorizing minds are insufficient when faced with contemporary developments (Oakes et al. 2015). Consequently, this situation necessitates that current educational approaches and practices are updated. Today, in most of the developed countries, individuals want to be provided with the twenty-first century knowledge and skills, which include the ability to rapidly access and implement new information, oral and written communication, critical thinking, problem solving, teamwork and collaboration, the ability to use technology, leadership and project management (Güneş 2014). Differences in the way new generation learners acquire their knowledge should be taken seriously (Breuer et al. 2017). For this reason, planned studies are conducted at every level from primary school to university and contemporary education approaches and methods are utilized (Güneş 2014).

In this context, it is evident that not only one method or technique should be used when teaching lessons at every educational stage, but different methods and techniques should be chosen appropriately according to the specific purposes. The educator should understand the students and act accordingly so that everyone can learn if the time and the opportunity are given to them (Petruţa 2013).

One of the easiest ways to achieve the goals that are stated in the curriculum is to use the right teaching methods and techniques, which should be chosen appropriately. Using a wide range of methods and techniques in class ensures that the students continue to be engaged and facilitates more effective learning. Educators' selection of methods is proportional to their ability to recognize and utilize very different methods (Romiszowski 2016).

No method can offer the perfect solution to a given course. Educators should choose the most appropriate methods for the class with their own personal efforts and they should be able to make changes based on their interpretation of signals from the classroom. The important thing is that the teacher should choose the method that will ensure that the subject is taught in the most effective manner (Nilson 2016). The effects of the methods and techniques vary according to the characteristics of the students, the subject matter, and goals that are desired to be achieved. For this reason, if the teachers have sufficient knowledge of the advantages and the disadvantages of different teaching methods and techniques, this will enable them to make the right decision when choosing the most appropriate method (Slunt and Giancarlo 2004).

In this study, we propose use of the Fuzzy PROMETHEE technique, one of the valuable multi-criteria decision-making techniques, for the analysis of the most widely used learning methods in education.

13.2 Theoretical Framework

13.2.1 Traditional Framework

Traditional teaching is a method that is led by the teacher and it uses methods such as direct instruction, question-answer and discussion. Traditional teaching is a teacher-centered model in which the teacher decides how to direct the course, the students, and how to make the assessment (Hannay and Newvine 2006). In traditional classrooms, the student is seen as an empty slate, and for this reason, it is essential that information is transferred effectively. There is an understanding that the information transferred by the teacher is adopted by the student as it is transmitted. The teaching process that determines what and how much the student has learnt is not taken into account in the traditional education method. In other words, in the traditional settings, the task of the student is to wait to be taught and the duty of the teacher is to convey the necessary information to the students (Thorne 2003).

In the traditional method, the teacher is seen as the only authority in the class. It is a teaching method that the students accept without questioning and in which interpretation, personal opinions and creative ideas are not encouraged. Individual differences between learners and learners' learning requirements are not considered. There is excessive dependency on textbooks. Students are not encouraged to perform research, and they do not make any effort to access the information. During the evaluation phase, the students return the information conveyed to them without adding any comments (Clayton et al. 2010).

It is stated in the literature that teaching methods based on the constructivist approach provide more effective learning than the traditional teaching methods (Muijs and Reynolds 2017; Schwerdt and Wuppermann 2011; Sriarunrasmee et al. 2015; Van Bergen and Parsell 2019).

13.2.2 The Collaborative Learning

Collaborative learning is a learning approach in which students form small groups in order to help each other learn an academic subject and in which the success of the group is rewarded in different ways (Slavin 2014). Along with group work in the collaborative learning process, the strategies applied and the problem-solving methods help students to see the differences between their own perspectives and the perspectives of the other students and it enables the students to learn many things from each other through decision-making and by effective cooperation. Since every student has their own responsibility in the collaborative learning process, there is unity among them, which enhances their social skills (Pattanpichet 2011). In this method, the teacher is a guide while the students are encouraged to be more active. In addition to teacher-student interaction, there is also student-student interaction. This allows the students to increase their socialization. Consequently, academic tasks that are given to a group are not only completed by one individuals, but the learners share their knowledge with each other (Gillies 2004).

The benefits of collaborative learning include: development of higher-level thinking skills, enhancement of student satisfaction with the learning experience, promotion of higher achievements and class attendance, and encouragement of student responsibility for learning (Barkley et al. 2014; López-Yáñez et al. 2015).

13.2.3   The Flipped Classroom

The flipped classroom is a model where the students can watch course videos that have been shared through the multimedia devices by the lecturer, read articles, think based on their prior knowledge and come to the classroom prepared, where the content will be discussed in greater depth (Zownorega 2013).

The main purpose of the flipped classroom model is to prepare the students for the topic before coming to class and consequently the quality of the face-to-face education in the classroom environment is enhanced (Bergmann and Sams 2014). The flipped classroom model of instruction is a relatively new teaching method which attempts to improve students' performance (Olakanmi 2017). In this method, the course contents, videos and the learning management systems are delivered to the students through multimedia devices. Students prepare questions on the topics they have not understood well. During the class, the activities that support learning such as finding answers to these questions together with their classmates, problem solving and discussing are carried out (Seaman and Gaines 2013).

There are many advantages of the flipped classroom method. The most important advantage is that it increases the teacher-student and student-student interaction during class time (Fulton 2012). Seamen and Gaines (2013) stated that the time spent on explaining the topic and revising in the classroom is transferred to the active learning activities because of the course video and students become actively involved in the process. According to Milman the main advantage of the flipped classroom method is that it supports teamwork within the classroom (Milman 2012). Other advantages of the model are that students can access the course videos whenever and wherever they want, and they have the opportunity to learn at their own pace (Kellinger 2012). Because each learner has different characteristics (Sweta and Lal 2017). In addition to all these advantages, Herreid and Schiller reported that the flipped classroom learning approach provides more time for students to conduct specific research (Herreid and Schiller 2013).

There are a number of studies in the literature that have found that the flipped classroom model increases the success of learners and positively affects their motivation and attitudes (Abeysekera and Dawson 2015; Chen et al. 2014; El-Banna et al. 2017; Wang 2017).

13.2.4 Gammified Flipped Classroom Merhod

The gamified flipped classroom is the usage of the gamification elements and game ideas in and out of the classroom in order to enhance the students motivation, success, their commitment to the education environment, enable them to solve problems they have encountered and enhance their learning (Aşıksoy 2017; Sırakaya 2017).

In order to implement an effective gamification design in the flipped classroom model, the gamification elements (dynamics, mechanics and components) are appropriately integrated into the teaching process after the educational attainments are determined (targets) (Bunchball 2010). The game dynamics found in the gamification strategy are: constraints, emotions, progression, relationships, and narrative. Mechanics are the basic processes that provide player participation and carry the action forward. Mechanics include challenges, chance, competition, cooperation feedback, rewards, transactions, turns, and win states. Components are more specific forms than those included in the mechanic and dynamics. It is possible to say that components serve certain aspects of the mechanics. Some examples of important game components are: achievements, avatars, badges, collections, gifting, leaderboards, points, virtual goods and levels (Werbach and Hunter 2012). It is clear from the studies analyzed regarding the flipped classroom model that it has more positive effects on the success and motivation of the students than the other models (Aşıksoy 2017; Hung 2015; Matsumoto 2016).

13.3 Data and Data Collection Methos

13.3.1 Participants and Experimental Design

A total of 180 first-year engineering students who were enrolled in a physics course were included in the study. The students were randomly assigned to Group 1 ($n = 45$), Group 2 ($n = 45$), Group 3 ($n = 45$), and Group 4 ($n = 45$). The Group 1 students learned with the traditional method, Group 2 students learned with the flipped classroom method, Group 3 students learned with collaborative learning and Group 4 students learned with the gamified flipped classroom method. Before the experimental procedure was initiated, each group was given instructions on how the courses were to be conducted. The content of the courses for each of the four groups was the same for 12 weeks.

13.3.2 Data Collecting Tools

Data were collected from a physics achievement test, science motivational questionnaire, physics attitude questionnaire and class attendance list.

13.3.3 Attendance List

Data about attendance were collected by an attendance list at each lecture, which was signed by all students who were present. In order to ensure that students did not sign for their absent friends, the number of attendants indicated by the list was cross-checked with the total number of students in the class. Each group attended two classes per week, each lasting 45 min.

13.3.4 The Physics Concept Test Attendance List

The physics concept test was used in order to determine the conceptual knowledge of students in all of the groups after the implementation process (after 12 weeks). The physics concept test consists of 34 questions and was prepared using examination questions from the previous years. Students marks were obtained from the physics concept test.

13.3.5 Physics Motivation Questionnaire

In order to identify the effect of the applied teaching methods on the motivation of the students, a SMQ (Science Motivation Questionnaire) developed by Glynn, Taasoobshirazi and Brickman (2016) was used. It consists of 22 items and the reliability coefficient of the motivation questionnaire was found to be 0.089. In the questionnaire, the minimum points a student could receive was 22, and the maximum was 110. High scores from the scale indicate high motivation towards the physics course.

13.3.6 Physics Attitute Questionnaire

To identify the effect of the related teaching methods on the attitudes of the students towards physics, a physics attitude questionnaire was used. It was developed by Pehlivan and Köseoglu (2011) consisted of 30 items. The reliability coefficient of the questionnaire was found to be 0.96. In the questionnaire, the minimum points a student could receive was 30 and the maximum was 150. High scores from the scale indicate a positive attitude towards the physics course.

13.3.7 Physics Self-efficacy Questionnaire

In order to identify the effect of the applied teaching methods on the self-efficacy of the students, the physics self-efficacy questionnaire developed by Tezer and Aşıksoy (2015) was used. It consists of 32 items, and the reliability coefficient of the self-efficacy questionnaire was 0.985. In the questionnaire, the minimum points a student could receive was 32, and the maximum was 160. High scores from the scale indicate high self-efficacy towards the physics course.

13.4 Fuzzy PROMETHEE Approach to Learning Models

The PROMETHEE method was developed in the 1980s by Brans, Vincke and Mareschal, (1986) and is based on a mutual comparison of each alternative pair with regard to each selected criteria. It is based on crisp data, which means it is not possible to use this method for vague conditions. Fuzzy logic was proposed by Zadeh (1965) in order to express vague conditions, experiences of experts or linguistic information mathematically. Fuzzy logic enables decision makers to simplify complex systems (Gravani et al. 2007). The Fuzzy PROMETHEE method is a hybrid method that is related to both fuzzy logic and the PROMETHEE method. This technique was proposed by Wang et al. (2008) for solving multi criteria decision making problems. The fuzzy PROMETHEE method allows the decision maker to use fuzzy input data, which gives more flexibility to the decision maker when comparing alternatives in vague conditions. Several studies have been conducted based on the fuzzy PROMETHEE (F-PROMETHEE) approach. Using this technique, Goumas and Lygerou (2000) ranked alternative energy exploitation projects, Bilsel et al. (2006) evaluated hospital web sites, Chou et al. (2007) evaluated suitable eco-technology method and Ozgen et al. (2011) applied this technique for the machine tool selection problem in a fuzzy environment.

This method enables the analysis of complex systems that have fuzzy parameters and produces effective comparison results. The details of this technique have been shown in the studies of Uzun Ozsahin et al. (2017a, b). They applied this technique in the evaluation of various types of nuclear medicine imaging devices and for the evaluation of cancer treatment techniques. Furthermore, Uzun Ozsahin and Ozsahin (2018) applied this technique in the analysis of breast cancer treatment techniques and Uzun Ozsahin et al. (2018) used it in the analysis of X-Ray based medical imaging devices. Maisaini et al. (2019), Ozsahin et al. (2019a, b, c), Sayan et al. (2019) are also have used F-PROMETHEE approach to rank the alternatives in selection problems of the medical and health sciences. In this study, we used the same method for the evaluation of various learning models.

Learning is a process that occurs in terms of cognitive abilities as much as in the affective sense (Berber et al. 2010). In the learning process, affective variables such as motivation, attitudes, and learning satisfaction need to be examined so that learners' diverse needs and interests can be better understood (Samimy 1994). The affective dimension of learning is important, not only because achieving a certain level of affective skills is important by itself, but it is sometimes critical in the process of acquiring the desired cognitive learning outcomes of education (Flake and Petway 2019; Lashari et al. 2012). Additionally, factors such as student-teacher relationships, use of technology, and student characteristics are variables that affect learning. Therefore, in our study, we considered both cognitive and affective features in the evaluation of learning methods.

In this application, we firstly collected all the necessary parameters for the most commonly used learning methods, as can be seen in Table 13.1. Since most of the parameters are not crisp data, we have shown those parameters with a fuzzy linguistic scale according to experts' opinions. Then, we applied the Yager index for the defuzzification of these fuzzy parameters and we applied the PROMETHEE technique with a Gauss preference function to determine the evaluation results. Then, we calculated the importance weight of these parameters equally. Lastly, we applied the visual PROMETHEE decision lab program and we obtained the complete ranking results of the alternative learning methods.

Table 13.2 shows the linguistic fuzzy scale which has been used for this application.

Table 13.1 The parameters of the alternative learning methods

	Classroom participation	Physics marks	Attitude towards physics	Motivation towards physics	Self-efficacy towards physics
The flipped-classroom meth.	H	H	H	VH	M
The traditional classroom meth	VL	L	VL	VL	L
Collaborative learning	M	M	L	H	M
Gamified flipped approach	H	VH	H	VH	H

Table 13.2 Linguistic fuzzy scale

Linguistic scale for evaluation	Triangular fuzzy scale
Very high (VH)	(0.75, 1, 1)
Important (H)	(0.50, 0.75, 1)
Important (H)	(0.50, 0.75, 1)
Medium (M)	(0.25, 0.50, 0.75)
Low (L)	(0, 0.25, 0.50)

13.5 Results and Discussion

The complete ranking of the alternative learning methods with regard to their linguistic fuzzy parameters can be seen in Table 13.3. Positive flow values show the strength of the alternatives, while negative flow shows the lack of the alternatives. The net flow provides the complete ranking.

These results show that the most effective learning methods are the gamified flipped classroom method, and the flipped classroom method because of their high parameters, such as employing technologies for information (see in Table 13.1), while the collaborative and traditional methods were the least effective learning models because of their low parameters.

The positive and negative properties of these learning models according to their parameters are shown in Fig. 13.1. Understanding the advantages and disadvantages of these models can provide valuable information to decision makers so they can select the best option according to their specific needs.

This study can be extended in order to analyze different academic disciplines and levels (Kiral and Uzun 2017; Uzun and Kıral 2017). In this context, we intend to use the Fuzzy PROMETHEE technique to evaluate learning models for primary and secondary schools in future studies. Additionally, by using the Fuzzy PROMETHEE

Table 13.3 Complete ranking of alternative learning methods

Rank	Learning models	Net Flow	Positive flow	Negative flow
1	Gamified flipped classroom model	0.1768	0.1768	0.0000
2	The flipped classroom model	0.1246	0.1321	0.0072
3	The collaborative learning	−0.0127	0.0497	0.0615
4	The traditional learning method	−0.2887	0.0000	0.2887

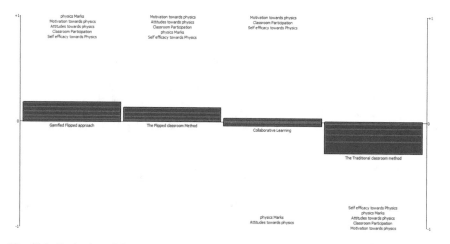

Fig. 13.1 Evaluation of the alternative learning methods

technique, the most appropriate models can be determined for teaching different academic disciplines.

The results of this study are beneficial for educators who want to choose appropriate teaching methods and strategies. However, the literature on fuzzy logic in the education field is limited. Therefore, this study can provide valuable guidance for future educational research.

13.6 Conclusion

Using multi-criteria decision making techniques such as PROMETHEE provides effective results by combining fuzzy information. Furthermore, while comparing with other multi criteria decision making techniques, it has been found that fuzzy PROMETHEE is proficient in numerous fields, since it depends on raw data. This method was used for analyzing the different learning methods and beneficial results were found and demonstrate. This study can be expanded by adding more criteria or alternative teaching methods. Both teachers and students can benefit extensively from the outcomes of this study and they can make comparisons between the different learning methods. Additionally, this technique can be applied to other decision making problems that arise in educational science.

References

Abeysekera L, Dawson P (2015) Motivation and cognitive load in the flipped classroom: definition, rationale and a call for research. High Educ Res Dev 34(1):1–14

Aşıksoy G (2017) The effects of the gamified flipped classroom environment (GFCE) on students' motivation, learning achievements and perception in a physics course. Qual Quant 1–17

Barkley E, Major C, Cross K (2014) Collaborative learning techniques. Jossey-Bass, San Francisco, California

Berber NC, Berber NC, Sarı M (2010) The effects of conceptual change-based teaching strategies on some affective feature improvements devoted to physics lesson. Ahi Evran Univ J Kırşehir Educ Faculty 11(2):45–64

Bergmann J, Sams A (2014) Flipping for mastery. Educ Leadership 71(4):24–29

Bilsel RU, Buyukozkan G, Ruan D (2006) A fuzzy preference ranking model for a quality evaluation of hospital web sites. Int J Intell Syst 21(11):1181–1197

Brans JP, Vincke P, Mareschal B (1986) How to select and how to rank projects: the PROMETHEE method. Eur J Oper Res 24:228–238

Breuer R, Sewilam H, Nacken H, Pyka C (2017) Exploring the application of a flood risk management Serious Game platform. Environ Earth Sci 76–93

Bunchball I (2010) Gamification 101: an introduction to the use of game dynamics to influence behavior. White Paper, 9

Chen Y, Wang Y, Kinshuk, Chen NS (2014) Is FLIP enough? Or should we use the FLIPPED model instead? Comput Educ 79:16–27

Chou WC, Lin WT, Lin CY (2007) Application of fuzzy theory and PROMETHEE technique to evaluate suitable eco-technology method: a case study in Shismen reservoir watershed. Ecol Eng 31(4):269–280

Clayton K, Blumberg F, Auld DP (2010) The relationship between motivation, learning strategies and choice of environment whether traditional or including an online component. Br J Edu Technol 41(3):349–364

El-Banna MM, Whitlow M, McNelis A (2017) Flipping around the classroom: accelerated bachelor of science in nursing students' satisfaction and achievement. Nurse Educ Today 56:41–46

Flake JK, Petway KT (2019) Methodologies for investigating and interpreting student-teacher rating incongruence in noncognitive assessment. Educ Measur Issues Pract 38(1):63–77

Fulton K (2012) Upside down and inside out: flip your classroom to improve student learning. Learn Lead Tech 39(8):12–17

Gillies RM (2004) The effects of cooperative learning on junior high school students during small group learning. Learn Instr 14(2):197–213

Glynn MS, Taasoobshirazi G, Brickman P (2016) Science motivation questionnaire: construct validation with nonscience majors. J Res Sci Teach 46(2):127–146

Goumas M, Lygerou V (2000) An extension of the PROMETHEE method for decision making in fuzzy environment: ranking of alternative energy exploitation projects. Eur J Oper Res 123(3):606–613

Gravani M, Hadjileontiadou S, Nikolaidou G, Hadjileontiadis L (2007) Professional learning: a fuzzy logic-based modelling approach. Learn Instr 17(2):235–252

Güneş F (2014) Öğretim ilke ve yöntemleri. Pegem A Yayınları, Ankara

Hannay M, Newvine T (2006) Perceptions of distance learning: a comparison of online and traditional learning. J Online Learn Teach 2(1):1–11

Herreid CF, Schiller NA (2013) Case studies and the flipped classroom. J College Sci Teach 42(5):62–66

Hung H (2015) Flipping the classroom for English language learners to foster active learning. Comput Assist Lang Learn 28(1):81–96

Kellinger JJ (2012) The flipside: concerns about the "new literacies" paths educators might take. Educ Forum 76(4):524–536

Kiral E, Uzun B (2017) Forecasting closing returns of Borsa Istanbul Index with Markov Chain process of fuzzy states. Pressacademia 4(1):15–24

Lashari TA, Alias M, Akasah ZA, Kesot MJ (2012) An affective-cognitive teaching and learning framework in engineering education. J Eng Educ 1(1):11–24

López-Yáñez I, Yáñez-Márquez C, Camacho-Nieto O, Aldape-Pérez M, Argüelles-Cruz AJ (2015) Collaborative learning in postgraduate level courses. Comput Hum Behav 51:938–944

Maisaini M, Uzun B, Ozsahin I, Uzun D (2019) Evaluating lung cancer treatment techniques using fuzzy PROMETHEE approach. Adv Intell Syst Comput 896:209–215

Matsumoto T (2016) The flipped classroom experience of gamified. Creat Educ 7(10):1475–1479

Milman N (2012) The flipped classroom strategy: what is it and how can it be used? Dist Learn 9(3):85–87

Muijs D, Reynolds D (2017) Effective teaching: evidence and practice

Nilson LB (2016) Teaching at its best: a research-based resource for college instructors (3rd edn). Wiley

Oakes J, Lipton M, Anderson L, Stillman J (2015) Teaching to change the world. Routledge

Olakanmi EE (2017) The effects of a flipped classroom model of instruction on students' performance and attitudes towards chemistry. J Sci Educ Technol 26(1):127–137

Ozgen A, Tuzkaya G, Tuzkaya UR, Ozgen DA (2011) Multi-criteria decision making approach for machine tool selection problem in a fuzzy environment. Int J Comput Intell Syst 4(4):431–445

Ozsahin DU, Uzun B, Musa MS, Şentürk N, Nurçin FV, Ozsahin I (2017a) Evaluating nuclear medicine imaging devices using fuzzy PROMETHEE method. Proc Comp Sci 120:699–705

Ozsahin DU, Uzun B, Musa MS, Helwan A, Wilson CN, Nurçin FV, Şentürk N, Ozsahin I (2017b) Evaluating cancer treatment alternatives using fuzzy PROMETHEE method. Int J Adv Comp Sci Appl 8(10)

Ozsahin DU, Ozsahin I (2018) A fuzzy PROMETHEE approach for breast cancer treatment techniques. Int J Med Res Health Sci 7(5):29–32

Ozsahin DU, Uzun B, Sani M, Ozsahin I (2018) Evaluating X-Ray based medical imaging devices with fuzzy preference ranking organization method for enrichment evaluations. Int J Adv Comput Sci Appl 9(3):7–10

Ozsahin DU, Nyakuwanikwa K, Wallace T, Ozsahin I (2019a) Evaluation and simulation of colon cancer treatment techniques with fuzzy PROMETHEE. IEEE Xplorer

Ozsahin I, Sharif T, Ozsahin DU, Uzun B (2019b) Evaluation of solid-state detectors in medical imaging with fuzzy PROMETHEE. J Instrum 14(01):C01019

Ozsahin I, Uzun B, Isa NA, Mok GSP, Ozsahin DU (2019c) Comparative Analysis of the common scintillation crystals used in nuclear medicine imaging devices. IEEE Xplorer

Pattanpichet F (2011) The effects of using collaborative learning to enhance students' English speaking achievement. J College Teach Learn (TLC) 8(11):1–10

Pehlivan H, Köseoglu P (2011) The reliability and validity study of the attitude scale for physics course. Proc Social Behav Sci 15:3338–3341

Petruţa GP (2013) Teacher's opinion on the use of interactive methods/techniques in lessons. Proc Social Behav Sci 76:649–653

Romiszowski AJ (2016) Designing instructional systems: decision making in course planning and curriculum design. Routledge

Samimy KK (1994) Teaching Japanese: consideration of learners' affective variables. Theor Pract 33(1):29–33

Sayan M, Sultanoglu N, Uzun B, Yildirim FS, Sanlidag T, Ozsahin DU (2019) Determination of post-exposure prophylaxis regimen in the prevention of potential pediatric HIV-1 infection by the multi-criteria decision making theory. IEEE Xplorer

Schwerdt G, Wuppermann AC (2011) Is traditional teaching really all that bad? A within-student between-subject approach. Econ Educ Rev 30(2):365–379

Seaman G, Gaines N (2013) Leveraging digital learning systems to flip classroom instruction. J Modern Teach Q 1:25–27

Sırakaya DA (2017) Student views on gamified flipped classroom model. Ondokuz Mayis Univ J Facul Educ 36(1):114–132

Slavin RE (2014) Cooperative learning and academic achievement: why does groupwork work? Anales de psicología/Ann Psych 30(3):785–791

Slunt KM, Giancarlo LC (2004) Student-centered learning: a comparison of two different methods of instruction. J Chem Educ 81(7):985–988

Sriarunrasmee J, Techataweewan W, Mebusaya RP (2015) Blended learning supporting self-directed learning and communication skills of Srinakharinwirot university's first year students. Proc Soc Behav Sci 197:1564–1569

Sweta S, Lal K (2017) Personalized adaptive learner model in e-learning system using FCM and fuzzy inference system. Int J Fuzzy Syst 19(4):1249–1260

Tezer M, Aşıksoy GY (2015) Engineering students' self-efficacy related to physics learning. J Baltic Sci Educ 14(3):311–326

Thorne K (2003) Blended learning: how to integrate online and traditional learning. Kogan Page Publishers

Van Bergen P, Parsell M (2019) Comparing radical, social and psychological constructivism in Australian higher education: a psycho-philosophical perspective. Austr Educ Res 46(1):41–58

Wang FH (2017) An exploration of online behaviour engagement and achievement in flipped classroom supported by learning management system. Comput Educ 114:79–91

Wang TC, Chen LY, Chen YH (2008, October) Applying fuzzy PROMETHEE method for evaluating IS outsourcing suppliers. In: 2008 Fifth international conference on fuzzy systems and knowledge discovery, FSKD'08, vol 3, pp 361–365. IEEE

Werbach K, Hunter D (2012) For the win: How game thinking can revolutionize your business. Wharton Digital Press, Philadelphia, PA

Zadeh LA (1965) Fuzzy sets. Inf Contr 8(3):338–353

Zownorega SJ (2013) Effectiveness of flipping the classroom in a honors level, mechanics-based physics class. Master Thesis, Eastern Illinois University, Charleston, Illinois

Chapter 14
The Most Common Factors Effecting Ground Water Quality

Maram Al Muhisen, Hüseyin Gökçekuş, and Dilber Uzun Ozsahin

Abstract From the presence of the first living forms on this planet, water has been an essential element for survival. Water is not only important for survival, but also for consumption, irrigation, manufacturing and other functions in various fields. Nonetheless, the waters have been continuously polluted by human activities. Such actions have made the waters unsanitary for consumption due to the inconsiderate waste dumping in the environment. Additionally, the increased population growth, construction developments, both industrial and economic expansions, and larger income costs are all factors that have influenced increased water consumption. The pressure placed on the water ecosystem has also increased extensively and qualitatively because of these factors. Moreover, the water ecosystem is faced with increasing possible risks due to the dumping of wastewater into areas surrounding the water sources from both households and industries. Seemingly, the use of improper irrigation methods may increase the amount of salinity and vaporization percentages in the water bodies on Earth. The following research will focus on the majority of ground water foundations, water complications, and the warm and mineral waters that have essentially affected our health due to its organic and harmful features. In addition, the most significant factors influencing the quality of groundwater will be

M. Al Muhisen (✉) · H. Gökçekuş
Faculty of Civil and Environmental Engineering, Near East University, Nicosia, Turkish Republic of Northern Cyprus, Turkey
e-mail: Maramalmuhisen@neu.edu.tr

H. Gökçekuş
e-mail: huseyin.gokcekus@neu.edu.tr

D. Uzun Ozsahin
DESAM Institute, Near East University, Nicosia, Turkish Republic of Northern Cyprus, Turkey

Department of Biomedical Engineering, Near East University, Nicosia, Turkish Republic of Northern Cyprus, Turkey

Medical Diagnostic Imaging Department, College of Health Science, University of Sharjah, Sharjah, United Arab Emirates

D. Uzun Ozsahin
e-mail: dilber.uzunozsahin@neu.edu.tr

© The Author(s), under exclusive license to Springer Nature Switzerland AG 2021
D. Uzun Ozsahin et al. (eds.), *Application of Multi-Criteria Decision Analysis in Environmental and Civil Engineering*, Professional Practice in Earth Sciences, https://doi.org/10.1007/978-3-030-64765-0_14

evaluated through the utilization of the Analytical Hierarchy Process (APH) method, which will concentrate on the following five factors; coal ash, manufacturing storage reservoirs, Arsenic, agriculture, ad seawater disturbance.

Keywords Polluted · Ecosystem · Wastewater · Water supplies · Ground water · Analytical hierarchy process

14.1 Introduction

Throughout the world, the issue surrounding the groundwater quality disputes has been increasing being that the pollution of subsurface water has turned into a prevalent issue (De Chaisemartin et al. 2017). Some of the issues on this matter include reports of pollution through coal ash (Lutey 2018), manufacturing storage containers (Baker 2018), seawater intrusion (Johnson 2018), arsenic (Pakianathan 2018), as well as cultivation (Bergquist 2018).

From outer space, the universe can be seen as a blue ball covered in water, with small and large islands scattered everywhere and as a result, the Earth has been referred to as the blue planet. The Earth is covered by a vast amount of water with a water body surface of approximately 71% and a volume of almost 1.973 billion cubic meters. Additionally, saline water accounts for approximately 97% of the overall water capacity found in oceans, seas, ponds, rivers and canals. The remaining 3% represents the freshwater, which is concentrated in some lakes, rivers and ponds. Additionally, the underground waters of the Arctic make up for the largest portion of fresh water that is available for human use, and representing about 1.6% of the total water capacity. Moreover, this percentage is inconsistent, especially with the inclining salinity ratio in the lakes, as well as the enclosed and semi-closed fresh water bodies, whose waters are linked to the saline seawater unilaterally (Mukherji and Shah 2005).

Due to the major existence of wide water bodies, there is a broad variety of water sources on Earth's surface, nevertheless, water can be classified according to its natural source as follows:

1. Ocean and sea waters
2. Rain water
3. River waters
4. Lakes/ponds
5. Groundwater
6. Mineral and hot waters

Controversially, scientists have classified the types of water based on its nature and components into two main types, and they are as follows;

1. Purified Water: this type of water is found on Earth's surface where it is instantly available for consumption, and in turn it is divided according to its salinity into:

   a. Saline Water; which contains high concentrations of evaporated mineral salts. Oceans and seas are the main source for salt-waters.

   b. Fresh Water; which contains low, or in some cases, no dissolved mineral salts concentrations. Rivers, streams, polar ice and rain are considered the main source of fresh water.

2. Subsurface Water: which is water that can be found underground, whether in saturated areas (areas that are filled with water) or unsaturated areas (areas that fall directly underneath the surface of earth). Its geological resources contains water and air within voids between the soil particles. This water type will be discussed in details later on.

It has been settled by various professionals that research on groundwater control establishments and strategies are required in order to determine the suitable prototypes needed for groundwater governance (Varady et al. 2010). Currently, intellectual pieces on groundwater control have achieved an increased attentiveness (Varady et al. 2010). Some of the prominent endeavors consisting of government representatives, scholars, as well as other proficient consultants entails the "Groundwater Governance: A Global Framework for Action"; which is a sponsored development by the GEF (Global Environment Facility), in addition to the Organization for Economic Cooperation and Developments WGI (Water Governance Initiative), whose concentrations are on controlling all forms of water. One researcher, Varady et al. observed ten research profiles on the groundwater control which exemplifies varied worldwide areas and local circumstances. Nonetheless, there has been a level of global awareness on groundwater approaches which involve various transboundary watersheds. Fundamentally, groundwaters are considered as local reserves for various excavations and management approaches. Studies involving subsurface water control and supervision all over the United States has been restricted. Mainly, studies have been turned towards watershed extent (Michaels and Kenney 2000), provincial range (Megdal et al. 2017), or transboundary aquifers (Sugg et al. 2015). However, various state-level subsurface water control and supervision studies have been implemented throughout the United States (Sugg et al. 2015), in which very little researchers have investigated water control and supervision on a state-level, even in the face of the reality that the majority of control mechanisms and supervision activities are concentrated on state-levels. Even so, researches associated to tackling significances and techniques on groundwater quality continue to be constrained.

14.2   Sources of Groundwaters

Groundwater is found in underground reservoirs, which is a rock or sedimentary layer that can hold a quantity of water and is composed of unstructured materials such as sand, pebbles or integrated rocks like sandstone or limestone, or they can be found in voids and cracks between soil granules (Wade 1987). There are several sources for groundwaters and they are as follows:

1. Rain water: The main source of groundwater is rain water, where part of this water is gathered along the earth's surface so that rivers may form, while some of it is filtered through the pores and cracks of the earth and gathered underground into a fixed reservoir form which is then transformed into water basins.
2. Mineral and Sulfur Water: Some lakes or rivers leak nearby, so the water is then gathered into basins underground and remains locked. These basins cannot be accessed or exploited except by drilling wells.
3. Magmatic Water: This type of water ascends the top following the various crystallization phases of the magma.
4. Connate Water: It is the water that accompanies the formation of sediments in the early stages and is trapped between their parts and pores (Mukherji and Shah 2005).

Groundwater is found at the top of the earth's crust, also known as the rocky silt area, and it is divided into two components:

1. Ventilation Range: It includes the higher portion of the rocky regions and the majority of rock voids are occupied with air and partially holds some water.
2. Saturation Range: It comes right after the ventilation range, towards the bottom, where the rock pores are occupied with water referred to as groundwater. The higher surface of the saturation range is referred to as (Water Table).

Modern science has been able to identify the amount of fresh groundwater throughout the universe, which is considered to be much larger than that available on the earth's surface. Groundwater is responsible for an estimated 98% of the world's total fresh water, with the exception of glaciers. While the fresh waters which are represented by fresh rivers and lakes, streams and the atmospheric clouds do not exceed 2%. Additionally, groundwater accounts for almost 0.6% of the total water on the planet, including fresh and saline waters. Notably, the groundwater may be renewable and flowing under the earth's surface while forming a network of sewers and rivers where the water maintains its levels, despite the constant consumption of the water, which is constantly recharged with rainwater that keeps falling or from the seepage of the rivers and lakes through the soil which accesses the groundwater. Contrarily, the ground water may not be renewable and could gradually decrease as it is consumed, where these waters are often below the surface which have accumulated underground in earlier centuries and rainy eras and lack any connections to renewable water sources. This type of groundwater has distinct characteristics from other forms of groundwater due to its presence in the ground for a long time, and these characteristics include high temperature and increase content of salts as well as dissolved gases, which is called hot mineral water. In some cases, it isn't necessary to excavate wells so that the groundwater can appear, it may explode in the form of fountains and springs as a result of increased pressure in the ground or crust pressure in the region. Water may flow from the fountain in the form of a waterfall due to the increased pressure applied onto the water or the pressure is reduced and the flowing water seeps onto the earth's surface and is drained into the watermills that crack and split waters. This water, which may be hot, derives its heat from the high height of the subsoil, or as a result of its proximity to places of volcanic activity, or is cold as a result of its emergence from layers close to the earth's surface.

14.3   Groundwater Control and Water Quality

The following segment will deliberate in what manner the nature of groundwater complicates control and supervision, in addition to the ideologies on efficient groundwater control and supervision. Although various paradigms are obtained from the United States, several annotations acquired in this section have become global.

14.3.1   The Nature of Groundwater and Control Complications

Most of the literary works regarding the matter of natural-pool reserves, advocate that public access may result in excess abuse of the reserves as well as being deprived of efficient establishments. Generally, people are incapable of refraining from utilizing natural pool reserves without the existence of a superficial authoritarian (Ostrom et al. 1999). With the presence of efficient laws and establishments that serve to restrict and identify inhabitant privileges, it is possible to prevent misuse as well as various harmful influences (Feeny et al. 1990).

Groundwater may be categorized as a natural-pool reserve as a result of many factors including; it's detracted capability; which means that every consumer is capable of minimizing the well-being of another consumer, in addition to its minimum excludability, which refers to access management (Schlager 2007). Groundwater can be exceptionally vulnerable to the issues linked to the natural aquifers, since they are comparatively economical and develop consistently at the instant availability of scientific knowledge and power to the probable consumers (Hou et al. 2015).

Complications linked to the quality of groundwater are remarkably demanding to alleviate as a result of the groundwater's natural-pool environment. As soon as pollution in groundwater takes place, it becomes very complicated to classify (Megdal et al. 2017) and treat (Varady et al. 2016). One researcher, Theesfeld, distinguish a group of attributes that cause complications in groundwater control. A variety of these attributes will be briefly discussed below.

1. Irreversibility; It should be known that consumption of groundwater can result in permanent physical destruction towards the aquifer or the terrain directly above from developments like ground collapse. Furthermore, the harm resulting from pollution can be costly, complex, or yet unfeasible for treatment, not even any mechanism can deliver a result like thrust and remediate or surfactant- enriched aquifer treatment (Kelly et al. 2013).

2. Time lag; Consequences following extraction or pollution require time to become evident. Therefore, time lags among extraction and succeeding effects provide water control with some challenges (Prakash and Datta 2014). The procedure for movement and transportation of groundwater take time. Following the activation of the sources, or in some cases once the source does not subsist, pollution can evidently be distinguished (Foster et al. 2013).

3. Indivisibility; It is not possible to confine or even physically shelter aquifers. However, the susceptibility of an aquifer relies on the form of pollution, extent of contamination, in addition to its hydrogeology (Kiparsky et al. 2017).

4. Hydrogeological vagueness; Due to the vast disparity in hydrogeology, in addition to the varieties of groundwater consumption, management and control becomes very complex. Such conditions can be witnessed in regions, like California, in which the ambiguities of aquifer peripheries and overlays in the GSA (Groundwater Sustainability Agency) limits multiple problems in groundwater control (Sugg et al. 2016). Surface-groundwater collaborations present difficulties for water control as a result of insufficient management between state organizations. Hydrogeology vagueness generates difficulties in obtaining the amount of water is contained by the transboundary aquifers in each region. Information is needed. Similarly, surface water material and groundwater material are generally of ambiguous trait. Generally, groundwater information is not as accessible as surface water information. It was advocated by Sugg et al. that for efficiency, groundwater requires more information material and technical assistances to serve the groundwater maintenance districts.

5. Construction of abstraction; There is insufficient monitoring not only on the quantity of wells that are extracted from groundwaters, but also on the amount of groundwater which wells abstract. For instance, Arizona was required to monitor the groundwater diminution so that the Central Arizona Project could be built (Garrick et al. 2011).

6. Material asymmetry; Typically, data on groundwater is restricted and distortedly alleged, resulting in complications for control. This happens in situations where water consumers hold better data on their historic water consumption techniques than the governing administrators (Patterson et al. 2017).

Moreover, size also causes complexities for control and supervision of groundwater. Specific aquifers, like the Ogallala Aquifer located in central United States, may lie behind thousands of square kilometers. Nevertheless, supervision and effects of groundwater consumption are explicit to their environment and in some cases location, given the disparity in geology and water consuming features in various sections of the aquifer (Patterson et al. 2017).

14.4 Analytical Hierarchy Process (AHP)

The analytic hierarchy process (AHP) is a structured methodology used for structuring and evaluating complicated decisions based on mathematics and psychology. The approach was introduced by Thomas L. Saaty during the 1970s, whom collaborated with Ernest Forman to establish the Expert Choice in 1983, which has been broadly studied and modified since. The methodology provides an accurate representation for quantifying weights of decision criterions. The process utilizes the experiences of individual experts in order to estimate the respective magnitudes of factors through pair-wise comparisons. Every respondent must compare the relative importance between two factors through a specially designed survey questionnaire. (Note that while most of the surveys adopted the five point Likert scale, AHP's questionnaire is 1–9 Li et al. 2019).

The analytical hierarchy process takes into account a set of evaluation criterions and a set of alternative options, in which, between the two sets, the best decision is selected. It should be noted, since some of the criteria may be contrasting, it is not generally true that the best alternative is that which optimizes every individual criterion. Rather, the alternative that attains the most suitable trade-off between the various criteria. For each evaluation criterion, the AHP calculates a weight based on the decision makers pair-wise comparisons for the criteria. The higher the calculated weight is, the more important the corresponding criterion is. As for a fixed criterion, the AHP appoints a score to each alternative matching the decision maker's pairwise comparisons of the alternatives based on that specific criterion. The higher the score, the better the performance of the alternatives of the corresponding criterion. Ultimately, the AHP incorporates the criteria weights and the alternative scores resulting in a global score for every alternative as well as a consequent ranking. The global score for a selected alternative is the weighted sum of the scores it acquired corresponding to all the criterions.

The AHP is a very adaptable and prevailing mechanism because the scores and final rankings are achieved based on the pair-wise comparisons of the criteria and alternatives presented by the user. The calculations produced by the AHP are always directed by the decision maker's knowledge. Therefore, the AHP can be considered as a mechanism capable of translating both qualitative and quantitative evaluations made by the decision maker into a multi-criteria ranking. Moreover, the AHP technique is straightforward because it does not require the structuring of a complex expert system with the decision maker's knowledge embedded within. Then again, the AHP may need a large quantity of assessments by the user, particularly for problems with various criteria and alternatives. Even though each individual assessment is effortless, the load of the assessment task can become unreasonable given that it only requires the decision maker to express how two alternatives or criteria compare to each other. Furthermore, the number of pairwise comparisons quadruples with respect to the number of criteria and alternatives. Take for example when comparing 10 alternatives on 4 criteria, $4 \cdot 3/2 = 6$ comparisons are required to construct the weight vector; and $4 \cdot (10 \cdot 9/2) = 180$ pairwise comparisons are required to construct the score matrix. The workload on decision makers can be reduced by partially or completely making the AHP automated, which is possible by identifying appropriate thresholds to decide the various pairwise comparisons automatically.

14.4.1 Applications and Functions

1. One form of AHP application is specifically within group decision-making (Saaty and Peniwati 2008). It is globally applied in a broad range of decision situations among fields such as government, corporate, manufacturing, healthcare, shipbuilding (Saracoglu 2013) and education.
2. Instead of presenting the "correct" decision, the AHP assists the decision makers to find one that best fits the objective and their comprehension of the problem. It offers a widespread and logical structure for constructing a decision problem, representing and quantifying its elements, linking those elements to overall objectives, and assessing alternative results.
3. Initially, when using the AHP, the decision-maker breaks down the problem into a hierarchy of more simply comprehended sub-problems that can be individually analyzed. The components of the hierarchy can correspond to any feature of the decision problem -tangible/intangible, measured carefully or roughly valued, well or vaguely comprehended- any factor that applies to the specified decision.
4. After the hierarchy is composed, the decision maker thoroughly analyzes the elements through the pairwise comparisons, which allows the decision maker to use concrete information regarding the elements. Generally, decision makers rely on their judgements regarding the elements relative importance and significance. The foundation of AHP relies on not only human judgment but also the underlying information for performing the assessments (Saaty 2008).
5. Once the elements are evaluated, the AHP converts the outcomes into numerical quantities, which may be processed and compared throughout the entire extent of the problem. For each element in the structure, a numerical weight or priority is developed that allows the comparison of diverse and often inadequate elements in a more rational and reliable manner. This potential distinguishes the AHP method from the other decision-making methods.
6. The final step of the AHP computes the numerical priorities for each alternative, which represents the relative ability of an alternative to achieve the decision objective, allowing a direct consideration of the various options.
7. Most firms provide computer softwares to help execute the process. Even though the AHP can be applied by individuals working on direct decisions, it is best effective where teams of people are working on complicated problems, particularly those with high risks, concerning human judgements and perspectives, whose solutions have long-term effects (Bhushan and Rai 2004). The approach has remarkable benefits when crucial elements are too complex to be quantified, compared, or when communication among team members are hindered by their differences (field, terminology, viewpoints).

14.4.2 Several Decision Situations Where the AHP Method May Be Applied Include

Choice: The process of selecting one option from a given set of options that is generally applied when the selection involves multi-decision criteria.

- Ranking: Structuring the set of alternatives from most desirable to least.
- Prioritizing: Establishing the competent elements in a set of alternatives rather than determining a single element or just simply ranking them.
- Resource Allocating: Distributing resources between a set of alternatives.
- Benchmarking: Comparing the organizations system with that of a well-known organization.
- Quality Management: Managing and handling the multi-dimensional features of quality and quality enhancement.
- Settling Conflicts: Resolving conflicts among individuals/groups with obvious opposing objectives.

The amount of AHP application in complicated decisions has numbered in the thousands (Steiguer et al. 2003) and developed remarkable outcomes in complications that involve planning, resource allocation, prioritizing and selection between alternatives (Bhushan and Rai 2004). Several fields that have also used AHP to solve problems are forecasting, total quality management, business process reengineer, quality function distribution and the balanced record. (Forman et al. 2001) Not all applications of AHP have been recorded, given that some of these applications occur in high-level organizations where privacy and security restricts the sharing of such information.

14.5 Methodology

The Analytic Hierarchy Process (AHP) is an approach that was first introduced by Saaty (Saaty 2008) to support multi-criteria decisions. The term 'Analytic' represents the process of breaking down the problem into fundamental elements. 'Hierarchy' represents the hierarchy structure of the fundamental elements that is listed with respect to the main objective. The term 'Process' represents the data and judgements that are processed to achieve the final outcome. The methodology has been broadly applied in numerous fields such as software selection complications (Min 1992), economic and management problem resolving (Yang and Lee 1997), plant location selection (Chan et al. 2004), supplier selection (Kahraman et al. 2003), assessment of project termination or maintenance based on the benchmarking technique (Liang 2003), choosing the best alternative among various outsourcing contracts relevant to maintenance services (Bertolini et al. 2004), and much more. The AHP is formed on two phases: (i) the hierarchy tree definition: (ii) the numerical assessment of the tree. The hierarchy tree definition begins at determining the proposed objective. The

Table 14.1 Saaty's fundamental scale

Intensity preference	Classification	Explanation
1	Equally favored	Activities C1 and C2 equally contribute to the purpose
3	Moderately favored	Experience and judgment somewhat prefer activity C1 to C2
5	Strongly favored	Experience and judgment strongly prefer activity C1 to C2
7	Very strongly favored	Activity C1 is very strongly favored over C2, its influence is evident in practice
9	Extremely favored	The evidence favoring activity C1 over C2, as the highest possible order of affirmation
2, 4, 6, and 8	Intermediate values	When a compromise is required

criteria and sub-criteria are then defined based on the expert's experience. Finally, the alternatives represent the leaves of the tree. The assessment phase is based on the pair-wise comparisons. The criteria that is found on the same level of the hierarchy are compared to determine the relative importance, which is compared to the criterion of the father-level. This process allows the user to (i) achieve values that weigh criteria, and (ii) define a ranking system for the alternatives. The assessment is bottom-up, where the decision-making procedure begins with the comparison of the alternatives with the criteria of the last level. The assessment continues to the criteria of the first level that is then compared to the objective. The scale shown in Table 14.1, was proposed by Saaty to rank the alternatives. This scale can be used to convert the linguistic judgments into numerical values. The AHP method incorporates the data to find a ranking for the alternatives; usually it is a normalized vector. Once the evaluation is complete, a sensitivity analysis is preformed to examine the outcomes of the variation for the weight of a specific criterion. Using the sensitivity analysis one can (i) compute the robustness of the solution and (ii) define the criteria that highly affects the final outcome.

14.6   Result and Discussion

Tables 14.2 and 14.3 provide the rankings of the factors that impact groundwater. From the table we can conclude that the most common factor affecting ground water is seawater intrusion, and is followed by coal-ash. The remaining factors following respectively are industrial storage tanks, arsenic and agriculture.

Table 14.2 The factors matrix scores

Factors	Seawater intrusion	Coal ash	Industrial storage tanks	Arsenic	Agriculture
Seawater intrusion	1	3	5	7	9
Coal ash	1/3	1	3	5	7
Industrial storage tanks	1/5	1/3	1	3	5
Arsenic	1/7	1/5	1/3	1	3
Agriculture	1/9	1/7	1/5	1/3	1

Table 14.3 Factors rankings

Value	Value (%)
1. Seawater intrusion	51.28
2. Coal ash	26.15
3. Industrial storage tanks	12.90
4. Arsenic	6.34
5. Agriculture	3.33

14.7 Conclusion

Formerly, project implementation mechanisms worldwide have been observing vital modifications as a result of various researches advocating that selecting the proficient implementing technique is able to reduce the projects time and cost quantities by a third pro rata. It has been proven that the AHP methodology in decision making is very efficient and is capable of choosing the most desired common factor which impacts groundwater. The results gathered from this study advocate that the most typical factor effecting groundwater is seawater intrusion and the respectively succeeding factor is coal ash.

Factors that have been affecting the quality of groundwater has become an increasing widespread subject, due to the essential need of groundwater for our well-being. Moreover, defining the factors that impact groundwater quality is considered to be a vital necessity, since it assists in easing the complications faced in the management of groundwater. Hence, the following study is capable of providing an insight on the methodology that can be utilized to help in understanding and defining the factors that impact the groundwater qualities.

References

Baker S (2018) Contaminated groundwater seeping into the trinity river from this spot needs costly fix

Bergquist L (2018) DNR board approves measure to limit manure pollution in Eastern Wisconsin to protect groundwater. Milwaukee J Sentinel, USA

Bertolini M, Bevilacqua M, Braglia M, Frosolini M (2004) An analytical method for maintenance outsourcing service selection. Int J Qual Reliab Manage 21(7):772–788

Bhushan N, Rai K (2004) Strategic decision making: applying the analytic hierarchy process. Springer, London. ISBN 978-1-85233-756-8

Chan AHS, Kwok WY, Duffy VG (2004) Using AHP for determining priority in a safety management system. Ind Manage Data Syst 104(5):430–445

Co-operation and Development

De Chaisemartin M, Varady RG, Megdal SB, Conti KI, van der Gun J, Merla A, Nijsten GJ, Scheibler F (2017) Addressing the groundwater governance challenge. Springer, Switzerland

de Steiguer JE, Duberstein J, Lopes V (October 2003) The analytic hierarchy process as a means for integrated watershed management (PDF). In Renard, Kenneth G (eds) First interagency conference on research on the watersheds. Benson, Arizona: U.S. Department of Agriculture, Agricultural Research Service, pp 736–740

Feeny D, Berkes F, McCay BJ, Acheson JM (1990) The tragedy of the commons: twenty-two years later. Human Ecol J 18:1–19

Forman, Ernest H, Gass SI (2001) The analytical hierarchy process—anexposition. Operat Res 49(4):469–487. https://doi.org/10.1287/opre.49.4.469.11231

Foster S, Hirata R, Andreo B (2013) The aquifer pollution vulnerability concept: aid or impediment in promoting groundwater protection? J Hydr 21:1389–1392

Garrick D, Lane-Miller C, McCoy AL (2011) Institutional innovations to govern environmental water in the Western United States: lessons for Australia's Murray-Darling. J Appl Econ Pol 30:167–184

Hou ZY, Lu WX, Chu HB, Luo JN (2015) Selecting parameter-optimized surrogate models in DNAPL contaminated aquifer remediation strategies. J Environ Eng Sci 32:1016–1026

Johnson J (2018) Farm Bureau declares opposition to proposed Salinas Valley New wells Moratorium

Kahraman C, Cebeci U, Ulukan Z (2003) Multi-criteria supplier selection using fuzzy AHP. Logist Inf Manag 16(6):382–394

Kelly BF, Timms WA, Andersen MS, McCallum AM, Blakers RS, Smith R, Rau GC, Badenhop A, Ludowici K, Acworth RI (2013) Aquifer heterogeneity and response time: the challenge for groundwater management. J Crop Pasture Sci 64:1141–1154

Kiparsky M, Milman A, Owen D, Fisher AT (2017) The importance of institutional design for distributed local-level governance of groundwater: the case of California's sustainable groundwater management act. J Water 9:755

Li et al (2019) Ranking of risks for existing and new building works. Int Sustain 10:2863

Liang W-Y (2003) The analytic hierarchy process in project evaluation. An R&D case study in Taiwan. Benchmarking: An Int J 10(5):445–56

Lutey T (2018) Cleanup of toxic coal ash that contaminated colstrip groundwater begins. Billings Gazette, USA

Megdal SB, Gerlak AK, Huang LY, Delano N, Varady RG (2017) Innovative approaches to collaborative groundwater governance in the United States: Environ

Michaels S, Kenney DS (2000) State approaches to watershed management: Transferring lessons between the Northeast and Southwest. In: Proceedings of the watershed management and operations management conferences. USA

Min H (1992) Selection of software: the analytic hierarchy process. Int J Phys Distrib Logist Manag 22(1)

Mukherji A, Shah T (2005) Groundwater socio-ecology and governance: a review of institutions and policies in selected countries. Hydr J 13:328–345

Ostrom E, Burger J, Field CB, Norgaard RB, Policansky D (1999) Revisiting the commons: local lessons. Global Chall 284:278–282

Pakianathan R (2018) Study measures arsenic contamination in wells. The Dartmouth, UK

Patterson L, Doyle M, Monsma D (2017) The future of groundwater: a report from the 2017 Aspen Nicholas water forum. The Aspen Institute, USA

Prakash O, Datta B (2014) Characterization of groundwater pollution sources with unknown release time history. J Water Resour 6:337–350

Saaty TL (June 2008) Relative measurement and its generalization in decision making: why pairwise comparisons are central in mathematics for the measurement of intangible factors— the analytic hierarchy/network process (PDF). Review of the Royal Academy of Exact, Physical and Natural Sciences, Series A: Mathematics (RACSAM). 102(2): 251–318. Cite-SeerX 10.1.1.455.3274. https://doi.org/10.1007/bf03191825. Retrieved 2008-12-22

Saaty TL, Peniwati K (2008) Group decision making: drawing out and reconciling differences. RWS Publications, Pittsburgh, Pennsylvania. ISBN 978-1-888603-08-8

Saracoglu BO (2013) Selecting industrial investment locations in master plans of countries. Eur J Industr Eng 7(4): 416–441. https://doi.org/10.1504/ejie.2013.055016

Schlager E (2007) In the agricultural groundwater revolution: opportunities and threats to development. Wallingford, UK

Sugg ZP, Varady RG, Gerlak AK, de Grenade R (2015) Transboundary groundwater governance in the guarani aquifer system: reflections from a survey of global and regional experts. J Water Int 40:377–400

Sugg ZP, Ziaja S, Schlager EC (2016) Conjunctive groundwater management to socio-ecological disturbances: a comparison of 4 western U.S States. Texas Water J 7:1–24

USA: Monterey Herald

USA: Sandra Baker

Varady RG, Weert F, Megdal SB, Gerlak A, Iskandar CA, House-Peters L (2010) Groundwater governance: a global framework for country action. GEF, USA

Varady RG, Zuniga-Teran AA, Gerlak AK, Megdal SB (2016) Modes and approaches of ground-water governance: a survey of lessons learned from selected cases across the globe. J Water (Switzerland) 8(10):417

Wade R (1987) The management of common-property resources: collective action as an alternative to privitisation or state regulation. Camb J Econ 11:95–106

Yang J, Lee H (1997) An AHP decision model for facility location selection. Facilities 15(9):241–254

Chapter 15
Selecting the Best Public–Private Partnership Contract by Using the Fuzzy Method

Maram Almuhisen, Huseyin Gökçekuş, Berna Uzun, and Dilber Uzun Ozsahin

Abstract Universally, many countries have been utilizing Public–Private Partnerships (PPPs) for exploiting infrastructure developments. Transactions in PPP, in such situations, are established on an association of complicated legitimate contracts. Conversely, in such transactions, a PPP contract is usually drawn up between the public authority (the "Contracting Authority") and private company (the "Private Partner") in the outline of a franchise contract or similar document. When drawing up and confirming PPP transactions, extensive time and investments are entailed due to the intricacy, difficulty, as well as the reality that the attributes of the infrastructure developments are often deeply consulted as to be reflected in such contracts. As a result, observers have questioned whether time and cost reduction in such procedures is viable, in which the provisions established in the franchise arrangements and other PPP indentures between the public and private sectors are systematized.

M. Almuhisen · H. Gökçekuş
Faculty of Civil and Environmental Engineering, Near East University, Nicosia, Turkish Republic of Northern Cyprus, Turkey
e-mail: maram.almuhisen@neu.edu.tr

H. Gökçekuş
e-mail: huseyin.gokcekus@neu.edu.tr

B. Uzun (✉) · D. Uzun Ozsahin
DESAM Institute, Near East University, Nicosia, Turkish Republic of Northern Cyprus, Turkey
e-mail: berna.uzun@neu.edu.tr

D. Uzun Ozsahin
e-mail: dilber.uzunozsahin@neu.edu.tr

B. Uzun
Department of Mathematics, Near East University, Nicosia, Turkish Republic of Northern Cyprus, Turkey

D. Uzun Ozsahin
Department of Biomedical Engineering, Near East University, Nicosia, Turkish Republic of Northern Cyprus, Turkey

Medical Diagnostic Imaging Department, College of Health Science, University of Sharjah, Sharjah, United Arab Emirates

Various regions have attempted to establish PPP indentures that are completely systematized in numerous types of infrastructure developments, such as road and railway networks, harbors or power generators. Thus far, for PPP transactions, no attempt has been made to develop a globally conventional linguistic on an international basis. Public–Private Partnership contracts can be categorized into various types, which are often known as PPP's or P3's. In this paper, six types of contracts will be discussed, namely Build-Operate-Transfer (BOT), Build-Own-Operate (BOO), Build-Own-Operate-Transfer (BOOT), Design-Build (D/B), Design-Build-Finance-Operate (DBFO), O&M (Operation and Maintenance), and their effectiveness will be ranked according to the selected parameters. This paper is aimed at establishing a fuzzy-based multicriteria decision-making model for determining which of the PPP contracts is the most suited for infrastructure projects.

Keywords Fuzzy · Multi-criteria decision making · Build-operate-transfer · Build-own-operate · Build-own-operate-transfer · Operation and maintenance

15.1 Introduction

Even now, the practice of project implementation undergoing significant transformation around the world. It has been implied through various studies and experience that selecting the most appropriate project delivery system may result in a decrease in the project duration and cost by up to thirty percent. Consequently, it can be claimed that the selection of project delivery system is a crucial strategic decision, which is achieved towards the end of practicality studies and are correspondent to the decision-making of the projects financial requirements approach. Therefore, researching and classifying the various project delivery systems is crucial in order to determine the most suitable system that conforms to the necessities of the manager and project (Thomas 2003). Around 1997, the term PPP began to be extensively applied by the new Labour government in the UK. They were looking for a new approach to infrastructure development, granting that the recent PPP program was founded on the preceding Private Finance Initiative (PFI), which was established under the departing Conservative Government. The PPP program offered a new labeling for PPP as a partnership based on contracts for public infrastructure between the public and private sectors (Public Private Partnerships 2001). Representation of the partnership by the PPP method between the public and private sectors has been expressed internationally with other terms including: Private Participation in Infrastructure (PPI) and Private-Sector Participation (PSP), which have been utilized in the development-finance division, P3, Privately-Financed Projects (PFP), P-P Partnership, which is an approach for evaluating currency exchange rates, and the Private Finance Initiative (PFI).

Correspondingly, the constructive attributes of PPP layouts in establishing infrastructure developments are deemed to be desirable for the Candidate Countries (CCs) of Central Europe due to the fact that such countries require massive financing, necessary effective public services, expanding market stability, and denationalization trends, which can establish a promising environment for private ventures.

Given that PPP layouts are available in many outlines and are yet a developing theory, they must be modified according to the specific necessities and attributes of each project and project partner. In order for a PPP to be productive, an efficient judicial and control structure is necessary; in addition, each partner is required to understand the purposes and demands of the other party. The European Commission has acknowledged the significance of PPPs and the necessity of an applicable authorized framework in order to guarantee that the rules and principles of the Treaty are employed. Subsequently, they have announced a statement to the Internal Market Council (Thomas 2003) among the tenders in order to alter the purchasing protocols and provide informational consultation on franchises. Such strategies have been established to assimilate significant documentations and are completely founded upon them (Sebastiaan and Menheere 1996).

There are many standards based on which PPP contracts are evaluated in terms of preference. The following standards were selected based on their commonness and significance and are classified as follows: financial strength, technical capability, management competence, correspondent experience, credit level, venture requirement, project integrity, social reputation, government guarantee, data clarity, contract dispositions, risk-taking, government reliability, competitive restriction, risk-sharing, interest fee guarantee, expense guarantee, legitimate guarantee, government funding, endorsement guarantee, and political funding. The financial strength, which assesses monetary reliability, includes investment strength, financial capability and financial assurance. Technical capability, on the other hand, is essential in developing project constraints, which involves the development of execution plans, innovative technology, vital equipment and alternate workforces. Management competence is crucial for achieving effective accomplishment of developments and incorporates valid administration, capability of project management, consultation and collaboration competence, franchise extent, aptitude of maintenance, and risk management potential. Correspondent experience is vital to decision making as is financial, constructional, and operational experience. Credit levels control the budget and time elements of the project (Aziz 2007). In this research the various types of PPP contracts will be discussed in detail. Furthermore, the PPP contracts will be ranked based on their evaluations obtained from the fuzzy-based Preference ranking organization for enrichment evaluation (PROMETHEE) approach.

15.2   Build–Operate–Transfer (BOT)

Due to the intensity and financial efficiency of the public economy, it was only valid to allow the infrastructure system to expand. Literally, in many countries, an escalating trend has appeared in which the governments have petitioned their private sectors to financially support public projects through investments. This trend is the outcome of two main causes, which are the deficiency of public funds as well as the noninterventionist tactic of the government organizations. Hence, in order for the government to be capable of subcontracting public developments to the private sectors, the Build Operate Transfer (BOT) methodology is preferred.

The Build Operate Transfer (BOT) approach was initially utilized In Turkey in an official private facility establishment in the year 1984by the Prime Minister Ozal. The establishment was initially a part of a vast denationalization program aimed at exploiting new infrastructure (Argyris and Kagiannas 2003). Nevertheless, this methodology was previously utilized in 1834 during the construction of the Suez Canal, whereby the canal was financed by European investment through Egyptian financial funding. Hence, the Egyptian ruler, Pasha Muhammad Ali, was appointed a franchise to design, construct, and operate the revenue producing waterway (Argyris and Kagiannas 2003).

One can define the Build Operate Transfer (BOT) as a methodology in which the private sector maintains a franchise from a public party, also known as the client, for a scheduled period of time in order to establish and operate a public facility. This establishment entails the funding, modelling and construction of the facility as well as administering and sustaining the facility sufficiently, including rendering it appropriately beneficial. The private sector procures profit return through managing the facility and during this time, the private sector portrays the owner. Once the concession period is completed, the private sector reassigns the proprietorship of the facility to the client exempt of liens and free of charge (Verhoeven 1997).

It has been found that in bidding circumstances, BOT projects are very beneficial. Moreover, such approaches, when executed, allow the company or the government to jointly share the project risks (Valencia 1997). Usually, BOT is utilized to establish a distinct resource instead of a whole system, which is usually modern or naturally undeveloped, in which renovation may be encompassed. Normally, the project company, which is manipulated by BOT, or the proprietor acquires its profits by charging the government with the payment's rather than charging the consumers with duties. In certain common law countries, various developments are known as concessions, such as highway toll developments, which are contemporary and are similar to BOTs.

In each BOT development, there are five main participants: the principal, the concessionaire, the investors, the contractor and the operator. Generally, the principal is the public sector, a government agency (local or federal) that comprehends the necessity for a public accommodation; however, they are incapable of monetarily funding the development. Throughout the concession period, the concessionaire is the owner of the development, and through utilization of the accommodation, apprehends the revenues on the initial investment. The investors include both stockholders and financiers, who offer financial assistance for the project. The construction of the project, including employing subcontractors, suppliers and consultants, is dependent upon the contractors. The operator is employed by the concessionaire and administers the operative phase of the accommodation (Menheere 1996).

The BOT has several advantages and disadvantages. Some of the most important advantages include the exploitation of the private sector assets rather than the public sectors, shifting technical expertise which is beneficial for developing regions, the placement of all risks onto the private sector, and finally, political resistance is minimized when utilizing private sectors due to the fact that at the end of the concession agreement, the project will be retransferred to government ownership (Kumaraswamy 2001). On the other hand, some of the disadvantages include that such developments are financially and technically complex, in addition to the requirements for superior professionals and consultants, as well as the enlarging expenses of consumers during the operating period and the conflict between the private and public sector profits.

15.3 Build-Own-Operate-Transfer (BOOT)

Build-Own-Operate-Transfer, known as BOOT, has many suitable factors that enable it to be an efficient approach used for project delivery by governments. It includes an established political structure, expectable and verified legal system, government support for a project that will benefit the community, long term petition, constrained opposition, rational revenues, beneficial cash flows, and foreseeable risk circumstances.

Build-Own-Operate-Transfer can be defined as a structure and type of concession established between the public sector and private sector where a certain portion of an infrastructure project, such as water, transportation, power, and telecom industries, are designed, built, owned and operated. In this structure, the concession grants the right to obtain profits from the project within a given time period of about 15–25 years, and shifts the ownership through a single establishment or association (BOOT provider) to the public sector afterwards (Arndt 1999). The income obtained can be dependent on various measures, which are efficiently two-part dues such as a fixed rate to the unit rate or the "Take-or-pay" agreements.

Participants in the BOOT methodology, which include the government, the investors, the contractor, the operator, the SPC (special purpose company) and the benefactors, are all required to minimize the investment expenses as well as the government's role in building, operating, and maintaining the infrastructure developments. Hence, this not only creates job opportunities for unemployed citizens, but also a liable environment for adequate and consistent quality, providing associates, as well as proposing advanced and alternative technology.

Build-Own-Operate-Transfer can be a beneficial approach for both the public and private sectors. Some advantages of this approach include the shared risks where construction and long-term operating risks are shifted to the BOOT operators, in addition to the various participants who are also involved in such developments. Moreover, through the incorporation of this method, project inevitability and initial revenue salvaging are possible and there is prominent liability for asset design. Correspondingly, BOOT allows operators to expand their expertise and knowledge on project management through experience. This methodology also involves intense financial enticements for the operator, reduced budgets of company constructing matters, construction and service delivery as a result of recovered expenditures, and finally, BOOT promotes extreme modernization allowing for the most competent designs.

Disadvantages are also associated with the BOOT approach, which include a larger budget for the end user as a result of the complete financial liability and continuing maintenance of the BOOT provider, undesirable feedback of private sector involvement from society, unattainable full benefits of the economic development, time and resource consumption of observing the operation contract of BOOT operators, and the necessity of a precise selection procedure for determining a BOOT partner (Jefferies 2002).

15.4 Build-Own-Operate (BOO)

The BOOT and BOO approaches have many similarities and dissimilarities in their concessions. The concession agreement in the BOO method does not shift possession back to the public sector or government once the arrangement is completed. A key variance between the two approaches is their financial support, where investors are only required to fund the project to its expected viable cashflows (Woodward 1995).

The Build Own Operate developments utilize certain financial structures, which are convoluted with respect to the numerous associations incorporated and the equivalent number of contracts where they all essentially interconnect (Confoy et al. 1999).

When employing the BOO approach, the private sector is obligated to design, build, finance, own, operate and maintain the infrastructure development during the concession phase. Generally, at the end of the concession phase, the BOO entails that the infrastructure development is transferred back to government ownership.

15.5 Design-Build-Finance-Operate (DBFO)

When Design Build Finance Operate (DBFO) contracts are utilized in projects, the private sectors offer resources and assembles debit financing from mercantile banks for a large portion of the asset cost and equity in order to balance the capital requirements and continuing management and maintenance services. Evidently, upon finalizing the project, the public sector pays for the asset as well as the services when granted. Once the project is completed, the private sector is compensated and the public sector pays an investment rate on the contract's duration period in order to pay off the banks and reimburse the equity. DBFO is a contract that fixated on productivity and establishes an efficient constraint. In such contract constraints, the public sector indicates the obligations, or in other terms "the what", and the private sector is left to establish and conclude the suitable manner, in order to convene the constraints. This contract raises the extent to which the private sector is able to modernize the designing solutions as to achieve the output constraints (British Highways Agency 1997).

Moreover, DBFO contracts have been capable of enhancing the initiation of cost effectiveness, advanced methods and a life-cycle assessment of the design and build of road plans, including the operation of the roads. Such enhancements are completely released when the private sector partakes in the layout design of the road plan. Hence, under a DBFO contract, the private sector is required to construct, operate and maintain such a layout plan in addition to taking responsibility for some of the planning risk. Most DBFO projects have proposed that the idea of planning risk will deliver better money value. Risk sharing within such contracts has been inciting, in which the protestor action and latent defect risk were the two areas that proved delivering of better money value to the private sector (British Highways Agency 1997).

15.6 Design–Build (D/B)

Unlike the previous contracts, a design-build contract is dependent on an individual accountability point. It focuses on decreasing the risk potential for the project owner in addition to minimizing the delivery agenda through corresponding the design and construction stage of the project. As a result of the individual accountability point, DB contracts incorporate clear fixes for the clients due to the fact that regardless of the nature of the responsibility, the contractor is accountable for all the work completed on the project (Murdoch and Hughes 2007).

Traditionally, construction projects usually appoint a designer on one side and a contractor on the other, but the DB alters the common work procedure. It grants the clients wish, an individual accountability point, in order to minimize risks as well as the total budgets. Presently, this approach is used in numerous countries and several forms of this type of contract are also available.

Occasionally, design-build is compared to the "master builder" methodology, which is considered as one of the oldest forms of construction sequence. While comparing the two innovative approaches, the writers of the Design-Build Contracting Handbook claimed that from a historical viewpoint, the traditional methodology is in fact an innovative theory as it has only been used for about 150 years. On the other hand, the "master builder" theory is said to have existed for more than four millennia (Kwak et al. 2009).

Conversely, the Design-Build Institute of America (DBIA) suggests that a contractor, a designer, a developer or a joint venture can lead a design-build based project, provided the entity maintains single contracts for design and construction (Robert and Michael 2001). Some architects have even proposed that architect-led design-build is a precise method to design-build.

15.7 O & M (Operation and Maintenance)

Building operation and maintenance, which is referred to as O&M, is defined as the continuous activity of preserving a building system's functioning corresponding to the objective of the design, the changing necessities of the owners and consumers, as well as the prime productivity stages. This procedure is capable of ensuring the preservation of the structure's overall cost-effectiveness by concentrating on occupant contentment, equipment dependability, and competent management. In operation and maintenance terms, competent management usually signifies activities that include arranging equipment and enhancing energy and comfort-control tactics in order to ensure that the equipment only functions to the extent required in order achieving its deliberate task. Furthermore, maintenance actions require physical examinations and caring for the equipment. When these tasks are methodically conducted, it results in increased reliability, minimization of equipment degradation and preservation of energy productivity.

With consumers expecting their suppliers to be more involved with the continuous operation of their equipment, the Operation and / or Maintenance (O&M) contracts have increased in popularity. Nevertheless, O&M contracts are extensive, and have been utilized in various situations, where each association has their own perspective on what is actually included in the contract. Such viewpoints can lead to misunderstanding, whether it is in the tendering phase or during implementation of the contract, hence provoking disagreements between the customer and contractor over accountability of certain elements of the O&M contract.

15.8 Methodology and Application

PROMETHEE is one of the effectively used multi-criteria decision-making methods that was originally developed by (Brans and Vincle 1985) and (Brans et al. 1986) in the late 1980s. When there is uncertain data about the parameters of the alternatives, the PROMETHEE technique is not applicable for the evaluation. In such cases, the-fuzzy based PROMETHEE method allows experts to describe these conditions using fuzzy input data. Recently, a number of different application of this hybrid technique have successfully been made by Geldermann et.al. 2000, Chou et. al. 2007 and Goumas and Lygerou 2000. Detailed information about the fuzzy based PROMETHEE technique has been presented in the study conducted by (Ozsahin et al. 2017). In order to process the data of the public private partnership contract, the triangular fuzzy scale has been used (see Table 15.1).

Table 15.1 Triangular fuzzy scale of the linguistic data

Linguistic scale	Triangular fuzzy scale
Very high (VH)	(0.75, 1, 1)
Importance (H)	(0.5, 0.75, 1)
Medium (M)	(0.25, 0.50, 0.75)
Low (L)	(0, 0.25, 0.5)
Very low (VL)	(0, 0, 0.25)

In this study, based on the selected criteria and their importance levels, the best public private partnership contract will be determined with the same methodology. The selected parameters of the public private partnership contract, including financial strength, capital strength, financing capacity, financial guarantee, technical ability, advanced technology, project implementation plans, personnel reserve, critical equipment, management ability, management formalization, project management capacity, ability of communication and cooperation, concession duration, risk management ability, maintenance ability, relevant experience, financing experience, constructing experience, operating experience, credit level, enterprise qualification, honor on projects, social reputation, government guarantee, legal guarantee, approval guarantee, government support, political support, interest rate guarantee, price guarantee, government credibility, government risk-sharing, restriction of competition, information transparency, contract spirits, and risk-taking, are given in Table 15.2 using a triangular fuzzy linguistic scale. Also, for the determination of the importance degree of the criteria, the linguistic fuzzy scale has been used. However, the Yager index is used for converting the triangular fuzzy numbers to a single point.

The most preferable alternatives are expected to have a maximum financial strength, maximum capital strength, maximum technical ability etc....Lastly, we used the visual PROMETHEE decision lab program with Gaussian preference function for each criterion given for ranking the PPPs alternatives. The net ranking results of the selected PPPs are presented in Table 15.3.

With a net flow of 0.0097, Design-Build-Finance is the most effective PPP contract and with a 0.0068 net flow, Operation and Maintenance contract is the second most effective PPP contract based on the given parameters and their importance degree.

In Fig. 15.1, the criteria that are above the zero value show the strength of the alternatives, while the criteria below the zero value show the weaknesses of the alternatives. These results provide very important information about the given PPP contracts. This result has been obtained with the given importance level of the parameters.

15.9   Conclusion

In this study, we proposed a fuzzy-based multi-criteria decision-making technique, PROMETHEE, for the analysis of public private partnership contracts. It provides researchers a good understanding about the advantages and disadvantages about the selected commonly used PPPs. The results showed that, Design-Build-Finance, Operation and Maintenance, Build–Own–Operate–Transfer are the top three most effective PPP contracts, while Build–Operate–Transfer, Design–Build, Build–Own–Operate contracts are the three least effective PPP contracts. Decision-makers could also obtain more sensitive ranking results by re-arranging the importance of the parameters, or by adding more parameters or alternatives.

Table 15.2 Data of the selected public private partnership contract

Criteria	Selected importance weight for criteria/aim	Build–operate–transfer (BOT)	Build–own–operate (BOO)	Build–own–operate–transfer (BOOT)	Design–build	Design–build–finance	O&M (operation and maintenance)
Financial strength	VH/Max	VH	VH	VH	L	L	H
Capital strength	M/Max	VH	H	H	L	H	M
Financing capacity	VH/Max	VH	L	VH	H	VH	H
Financial guarantee	H/Max	VL	H	L	VH	L	VH
Technical ability	VH/Max	H	VH	H	VH	M	VH
Advanced technology	VH/Max	L	M	M	L	L	H
Project implementation plans	H/Max	VL	M	M	VH	M	VL
Personnel reserve	VL/Max	VH	VH	H	H	L	H
Critical equipment	H/Max	VL	L	VH	L	VH	M
Management ability	H/Max	M	L	VH	H	H	M
Management formalization	VH/Max	VH	M	L	VH	M	VH

(continued)

Table 15.2 (continued)

Criteria	Selected importance weight for criteria/aim	Build–operate–transfer (BOT)	Build—own—operate (BOO)	Build–own–operate–transfer (BOOT)	Design–build	Design–build–finance	O&M (operation and maintenance)
Project management capacity	VH/Max	VL	VH	M	H	M	VH
Ability of communication and cooperation	VH/Max	M	VH	H	L	M	VH
Concession duration	VH/Max	VH	VH	H	M	VH	L
Risk management ability	VH/Max	M	L	H	VL	M	L
Maintenance ability	H/Max	M	VH	M	VL	M	VH
Relevant experience	H/Max	VH	L	VH	VH	VL	VL
Financing experience	VH/Max	H	L	VH	VH	H	VH
Constructing experience	VH/Max	M	VH	L	H	H	L
Operating experience	VH/Max	VH	M	VH	L	M	M
Credit level	H/Max	M	VL	M	H	VH	VH

(continued)

Table 15.2 (continued)

Criteria	Selected importance weight for criteria/aim	Build–operate–transfer (BOT)	Build—own—operate (BOO)	Build–own–operate–transfer (BOOT)	Design–build	Design–build–finance	O&M (operation and maintenance)
Enterprise qualification	H/Max	L	VL	M	L	H	H
Honor on projects	H/Max	H	H	M	L	H	VL
Social reputation	VH/Max	H	L	VH	H	H	H
Government guarantee Legal guarantee	VH/Max	VH	M	L	H	VH	H
Approval guarantee	VH/Max	H	H	M	VH	H	VL
Government support	H/Max	VL	L	M	L	VH	VH
Political support	VH/Max	H	M	M	H	L	H
Interest rate guarantee	VH/Max	VL	M	VH	VH	L	L
Price guarantee	VH/Max	VH	VH	L	L	H	L
Government credibility	VH/Max	M	L	H	H	L	VH
Government risk-sharing	H/Max	VH	VH	H	VH	VH	H

(continued)

Table 15.2 (continued)

Criteria	Selected importance weight for criteria/aim	Build–operate–transfer (BOT)	Build—own—operate (BOO)	Build–own–operate–transfer (BOOT)	Design–build	Design–build–finance	O&M (operation and maintenance)
Restriction of competition	VH/Max	M	H	VH	M	H	H
Information transparency	VH/Max	M	H	Fair	M	H	M
Contract spirits	H/Max	VH	H	VL	VH	VH	H
Risk-taking	H/Max	H	VH	VL	H	H	VH

Table 15.3 Complete ranking of the selected public private partnership contract

Ranking	Water	Positive outranking flow	Negative outranking flow	Net flow
1	Design–build–finance	0.0738	0.0641	0.0097
2	Operation and maintenance	0.0813	0.0744	0.0068
3	Build–own–operate–transfer	0.0809	0.0807	0.0002
4	Build–operate–transfer	0.0716	0.0749	−0.0032
5	Design–build	0.0760	0.0811	−0.0051
6	Build–own–operate	0.0750	0.0835	−0.0085

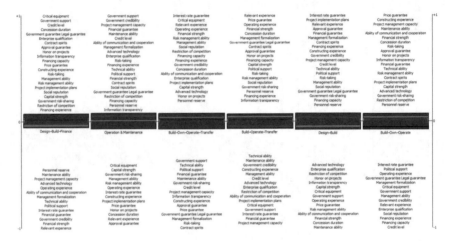

Fig. 15.1 PROMETHEE evaluation result of the selected PPPs

References

Argyris G, Kagiannas D (2003) The role of build operate transfer in promoting. National Technical University of Athens, Greece

Arndt H (1999) Is build-own-operate-transfer a solution to local government's infrastructure funding problems? Victoria. The University of Melbourne, Australia

Aziz A (2007) Successful delivery of Public-Private Partnerships for infrastructure development. J Constr 133:918–931

Brans JP, Vincle P (1985) A preference ranking organization method. Manag Sci 31(6):647–656

Brans JP, Vincke P, Mareschal B (1986) How to select and how to rank projects: the PROMETHEE method. Eur J Oper Res 24:228–238

British Highways Agency (1997) Value in roads. HIAG J02–7650. Stationary Office, London

Chou WC, Lin WT, Lin CY (2007) Application of fuzzy theory and PROMETHEE technique to evaluate suitable eco-technology method: a case study in Shismen reservoir watershed. Ecol Eng 31:269–280

Confoy B, Love D, Wood M, Picken H (1999) Build-own-operate—the procurement of correctional services in profitable partnering in construction procurement, London, CIB W92 and CIB TG 23 Joint Symposium

Geldermann J, Spengler T, Rentz O (2000) Fuzzy outranking for environmental assessment. Case study: iron and steel making industry. Fuzzy Sets Syst 115(1):45–65

Goumas M, Lygerou V (2000) An extension of the PROMETHEE method for decision making in fuzzy environment: ranking of alternative energy exploitation projects. Eur J Oper Res 123:606–613

Jefferies G (2002) Critical success factors of the BOOT procurement system: reflections from the Stadium Australia case study. Eng Constr Architect Manag 9(4):352–361

Kumaraswamy MZ (2001) Governmental role in BOT-led infrastructure development. Int J Proj Manag 195–205

Kwak Y, Chih Y, Ibbs W (2009) Towards a comprehensive understanding of public private partnerships for infrastructure development. Calif Manag Rev 51:51–78

Menheere P (1996) Case studies on build, operation, and transfer. University of Technology, Netherlands

Murdoch J, Hughes W (2007) Construction contracts: law and management. Taylor & Francis E-library, United Kingdom

Ozsahin D, Uzun B, Musa M, Şentürk N, Nurçin F, Ozsahin I (2017) Evaluating nuclear medicine imaging devices using fuzzy PROMETHEE method. Proced Comput Sci 120:699–705

Public Private Partnerships (2001) Introduction. Handbook, recommendations and conclusions. USA: Private Operations and Financing of TEN's

Robert C, Michael C (2001) Design-build contracting handbook. Aspen Law & Business, USA

Sebastiaan M, Menheere P (1996) Case studies on build operate transfer. Delft University of Technology, Netherland

Thomas M (2003) Measuring the impacts of delivery system on project performance. Construction Industry Institute, USA

Valencia L (1997) Financing methods in project management and its relevance in succeed of a project. University of Technology, Netherlands

Verhoeven L (1997) BOT in Netherlands. University of Technology, Netherlands

Woodward G (1995) Use of sensitivity analysis in build-operate-transfer project evaluation. Int J Project Manage 13(4):239–246

Chapter 16
Comparative Analysis for Irrigation Water Application Methods Using TOPSIS

Mukhtar Nuhu Yahya, Ibrahim Muntaqa Tijjani Usman, Hüseyin Gökçekuş, Dilber Uzun Ozsahin, and Berna Uzun

Abstract For a proper design of any irrigation system, an appropriate irrigation water application method is needed in order to select the best method among many. To come off with a better implementation and decision making, A lot of factors such as initial cost of installations, topography of the location, space, crop to be planted and many other factors are needed before any implementation and decision is made on the selection process. Thus, for the ranking and evaluation of different irrigation water application methods in this paper, Technique for Order of Preference by Similarity to Ideal Solution (TOPSIS) a widely used qualified multi-criteria decision making technique is used for the selection and will effectively identify the most appropriate

M. N. Yahya (✉) · H. Gökçekuş
Faculty of Civil and Environmental Engineering, Near East University, Nicosia, Turkish Republic of Northern Cyprus, Turkey
e-mail: mnyahya.age@buk.edu.ng

H. Gökçekuş
e-mail: huseyin.gokcekus@neu.edu.tr

M. N. Yahya · I. M. T. Usman
Department of Agricultural and Environmental Engineering, Faculty of Engineering, Bayero University, Kano, Nigeria
e-mail: imtusman.age@buk.edu.ng

D. Uzun Ozsahin · B. Uzun
DESAM Institute, Near East University, Nicosia, Turkish Republic of Northern Cyprus, Turkey
e-mail: dilber.uzunozsahin@neu.edu.tr

B. Uzun
e-mail: berna.uzun@neu.edu.tr

D. Uzun Ozsahin
Department of Biomedical Engineering, Near East University, Nicosia, Turkish Republic of Northern Cyprus, Turkey

Medical Diagnostic Imaging Department, College of Health Science, University of Sharjah, Sharjah, United Arab Emirates

B. Uzun
Department of Mathematics, Near East University, Nicosia, Turkish Republic of Northern Cyprus, Turkey

157

method in terms of cost, energy usage, maintenance and may more. To overcome the difficulty faced in the selection process by the decision makers, TOPSIS is proposed to deal with the vagueness in the decision makers' judgement. This study will use the TOPSIS to analyze the irrigation water application methods based on many criteria, and the outcome of the study will be used by the concerned parties in the field of irrigation in choosing the best among all the methods and will give an insight for the relevant stakeholders in the field of engineering, government and non-governmental organisations in decision making. From the result obtained in this study, basin method has been the most preferred method followed by border method for all the irrigation water application methods.

Keywords Irrigation systems · Multi-criteria decision making (MCDM) · TOPSIS

16.1 Introduction

The goal of irrigation practice is to supply water uniformly to crops in a way that each crop receives the amount of water it needs to enable optimum crop yield. Several techniques are used depending on the water availability, crop type and farmers' financial status (Bjorneberg 2013). Even so, the success of irrigation practice depends on the selection of the most appropriate method with respect to the environmental conditions, planning, design and installation, and operation of the system for the selected method (Arora 2014).

The properties of the root zone below the land surface which is mostly from 1 to 2 m of most plants, is the main interest of irrigation engineers and agriculturists (Arora 2014). However, to determine the infiltration characteristics of a soil, investigating layers of soil below the root zone is equally important because it indirectly affects the plants. Timing of water application and the quality and quantity of water to be applied are the main concerns in irrigation practice. The problem of timing of water application involves the computation of available moisture in the soil and the rate at which it is depleted. This is because the soil is only capable of storing a certain amount of water. However, water must be applied before the moisture in the soil is depleted to the level that plants are not able to tap water from the soil. The problem of the amount of water and its quality involves the computation of the quantity of water which will restore the soil's moisture to the required level, and analysis of the quality of water to prevent contamination of the soil and the plants (Kumar and Vishal Kumar 2012).

This chapter focuses on the comparative analysis of Irrigation water application methods using a decision-making tool, TOPSIS.

There are many criteria for the selections of various methods of irrigation water application, among these criteria are,

- Costs of the system and its appurtenances,
- Field sizes and shapes,
- Soil intake and water holding characteristics,

- The quality and availability (timing of deliveries, amount, and duration of delivery) of the water supply,
- Energy requirement,
- Climate,
- Cropping patterns,
- Historical practices and preferences, and
- Accessibility to precision land levelling services (topography) (NRCS 2012).

These criteria and some others are necessary for the selection of the best method of irrigation water application.

Irrigation water application methods are sub-categorised into two, three(or four in some text) to include conventional and advanced methods; surface and pressure irrigation methods; surface, sprinkler and sub-surface methods; and surface, sub-surface, sprinkler and Drip or trickle irrigation method (NRCS 2012; Kumar and Vishal Kumar 2012; Bjorneberg 2013; Arora 2014). However, within the context of this chapter, all methods will be considered as irrigation methods, irrespective of their sub-category. These methods include: Basin method, border method, furrow method, wild flooding method, sprinkler method, and drip or trickle method.

16.2 Irrigation Water Application Methods

16.2.1 Basin Irrigation Method

As the name implies, a basin is formed around the field to be irrigated, then water is pumped into the basin. This is the most common form of surface irrigation method, particularly in regions with small layout of fields (Khanna and Malano 2006). The basin method is a special form of check flooding used mostly for the irrigation of orchards. The check basin can be a square or a circular basin, (most of the time irregular in shape), the circular basin is referred to as the ring basin. Generally, each tree a given its own ring basin. However, if the conditions of soil and surface slopes are favorable, a basin can contain 2–4 trees. A basin may be furrowed or corrugated (depending on the soil type), it may have some raised beds to benefit some crop types, but as long as the inflow of water is not controlled and has no direction, no matter the modifications it remains a basin (NRCS 2012).

Not all crops and soil types are suitable for basin irrigation method. However, moderately, and slow intake soils, deep rooted, and closely spaced crops are favorable (Khanna and Malano 2006; NRCS 2012). Provision of surface drainage for runoff is not necessary, and reclamation of salt affected soil is easily achieved in basin irrigated fields (Bjorneberg 2013).

Even though the basin irrigation method is the most practiced method, with its low tech, land leveling is particularly important to achieve the desired uniformity and efficiency. Studies from different countries where this method is practiced have shown that, there is an overall low efficiency in irrigation due to over irrigation and under irrigation, which has become a norm in most countries (Miao et al. 2018). Water flow rate should be large enough to cover entire basin but not too much to over saturate the basin's soil root zone and to cause water spill over the basin's dike. The basin method is usually practiced in the irrigation of rice and fodder crops. Some vegetables like onions are also irrigated using the basin method (Kumar and Vishal Kumar 2012).

16.2.2 Border Irrigation Method

The border method can be regarded as an optimized basin irrigation method. With the introduction of slope, a specific shape, and free draining condition at the lower end, the border irrigation method is more efficient, and inflow has a control and direction (Bjorneberg 2013). In the border method, the land is divided into a series of strips. The strips are about 3–30 m wide and 100–800 m long, separated by low levees (or border or dikes). These strips have uniform slope (less than 0.5%) along the direction of flow, but there is no cross-slope. Sloping border are suitable for most crops except those that need prolonged pounding. Therefore, when the water is applied, it flows over the entire width as it moves down the slope along the longitudinal direction. Water is supplied to each strip from a ditch made of earth (or concrete). Water from the earth ditch is usually supplied to the strip by opening a path in the ditch bank. For concrete ditches, water is supplied to the strip using a gated openings or siphons made from suitable materials (e.g. plastic pipes) (Kumar and Vishal Kumar 2012). For groundwater sources, underground concrete pipes are used instead of supply ditches. As the water is discharged, it flows along the strip towards the lower end in the form of sheet. It is confined between levees. It infiltrates the soil as it flows. When the flow approaches the lower end, the supply to the strip is stopped. On a relatively leveled field, the strips are laid along the general slope of the field. In this case, the strips are straight and parallel to one another. On a relatively steep field, the strips are aligned along the contours, with their longer sides parallel to the contour. In other case, the strips are curved, these strips are known as contour strips (Zerihun et al. 2013).

In border method, the precision of the field topography is critical, although the extended length of strips allows the use of machineries for better leveling.

Border methods are grouped into three major categories: (fixed flow, cut back, and tail water reuse) depending on the management strategy adopted.

Generally, field efficiency ranges from good to excellent depending on the design and installation of the border strips and good water management. With a slope range of 0.001–0.002, water efficiency of 70–75% can be achieved on silty clay to clay soil with water application depth of 75–100 mm. For higher efficiency, stream size and flow must be controlled to match the moisture depletion rate of the soil to obtain approximately corresponding infiltration rate at both upper and lower ends (Arora 2014).

16.2.2.1   Design Criteria

The following are the general design criteria for the strips:

(1) Width
(2) Slope
(3) Length
(4) Location of levees and strips
(5) Computation of time to cover a strip (Arora 2014).

16.2.3   Furrow Irrigation Method

In furrow irrigation method water is applied to the field to be irrigated along a series of long narrow path called furrows. This is done to avoid spreading the whole land with water for minimization of water. In this type of surface irrigation, the furrows are dug at intervals and at right-angles to the field channels or laterals. Water application is dependent on the length of the furrows, water management factors and soil properties in furrow type of irrigation. Water is being flowed into the furrows for 12–24 h during irrigation exercise, and it can be longer or shorter depending on the aforementioned factors. It is worth noting that, water is not spread over the whole irrigated field. The water flowing in the furrows infiltrates the soil and spreads laterally and reaches the roots of the plants between the furrows. As discussed earlier in furrow type of irrigation only some portion of the land is wetted (i.e. in between the laterals), whereas in basin and border methods the entire field is wetted directly by the water (Bjorneberg 2013).

Furrows method provides better on-farm water management flexibility under many surface irrigation conditions. This method provides the irrigator opportunity to manage irrigation water toward higher efficiencies as irrigation field conditions changes throughout a season. However, the furrow method does not enjoy much higher application efficiencies than borders and basins. Additional disadvantages with furrow irrigation method may include:

(1) Salt accumulation problems are encountered between furrows.
(2) Increased level of tail water losses.
(3) Movement of farm equipments across the furrows is tedious.
(4) Added expense and time to make extra tillage practice (furrow construction).
(5) Increase in the erosive potential of the flow.
(6) Higher labour and operational demand.
(7) Generally, furrow systems are more difficult to automate, particularly about regulating an equal discharge in each furrow (NRCS 2012)

Ranging from 1/5 to 1/2 of the field, this method helps in reducing the evaporation losses and achieving high water-application efficiency. Water is supplied from the field channel (or a supply ditch) to the furrows either through small cut made in the banks of the field channels or through small portable siphons sometimes concrete pipes are placed underground at the upper (higher) end of the furrows instead of field channels. Supply gates are sometimes used at the point of flow. The gates are of small sizes, easily adjustable and are quite convenient for controlling the supply in the furrows. A collection basin is usually provided at the tail ends of the furrows to collect water for reuse.

Like border strips, the furrow method is classified into two based on the topographical nature of the field: Straight furrows and contour furrows (Kumar and Vishal Kumar 2012; Arora 2014).

16.2.3.1 Design Consideration

(1) Width and depth of furrows
(2) Length of furrows
(3) Slope of furrows
(4) Spacing of furrows
(5) Discharge (Kumar and Vishal Kumar 2012; Arora 2014).

16.2.4 Sprinkler Irrigation Method

In sprinkler irrigation (sometimes called overhead irrigation), water is supplied to the field from the main source through the sub-main, the laterals, the risers and out of the sprinkler head, all with the help of additional pressure ($1–4$ kg/cm^2) (Kumar and Vishal Kumar 2012). This method resembles light rainfall. It is the most aesthetic of all irrigation systems. Sprinkler irrigation can be used for almost all crops and most soils (NRCS 2016). For soil with very low infiltration rate, sprinkler systems are not advised. Sprinkler systems are particularly suitable for sandy soils and irregular topography. A professionally designed and managed sprinkler system results in a higher water efficiency, better water management, and increased crop yield. Product quality is also increased and about 15% of more land is available compared to surface irrigation methods. However, the initial investment in sprinkler system is exceptionally high compared to any surface irrigation method. Notwithstanding, in places of high wind, the water distribution is not uniform and evaporation increases (Phocaides 2007).

Sprinkler systems are of types: based on method of application, the sprinkler is divided into (i) fixed nozzles, (ii) perforated type, and (iii) rotating nozzles. The fixed nozzles are the oldest, where the nozzle is fixed on a vertical pipe called the riser and sprays in one direction. The perforated type is a pipe with number of small holes drilled along the side of the pipe with a pressure between 0.8 and 3 kg/cm^2. The rotating nozzles are the most recently used with a pressure range of $2–7$ kg/cm^2. For sprinkler system based on portability, these include, (i) mobile sprinkler systems which can be completely moved around the field or from one field to another. (ii) Semi-mobile system has its main line completely buried underground and the lateral lines can be moved from place to place. (iii) permanent system is when both main and lateral lines are buried underground, hence requiring less maintenance (Kumar and Vishal Kumar 2012; NRCS 2016).

A sprinkler irrigation system consists of the following main components:

1. Pumping unit
2. Main pipeline (or main)
3. sub-main pipeline (or sub-main)
4. Lateral pipelines (or laterals)
5. Vertical pipes or risers
6. Sprinkler head (or nozzle) (Phocaides 2007).

16.2.5 Drip Irrigation Method

In this system, water is emitted onto the soil near the roots or directly to the root zone of the plants through a special outlet device called an emitter or a dripper. The emitters have drip nozzles which supply the water drop after drop at a low rate, varying from 2 to 10 L per hour (Arora 2014; NRCS 2013). Water is supplied to the root zone just sufficient to maintain the soil moisture within the required crop growth range. Soil characteristics and water flow rate are factors that determine the area to be covered by one emitter. Depending on the water requirements and number of crops, numbers of emitters are designed to achieve irrigation target. Despite being the most efficient irrigation method, drip irrigation methods are affected by to major factors. Constant factors, which include: Climate and weather, soil type and characteristics, slope and environmental quality. For variable factors, water price and availability, water quality, energy, labor cost and availability, system quality and cost, crop type and quality, operation and maintenance, education of the irrigator, interest rate, depreciation rate, etc. can also be factors affecting the performance and efficiency of a drip irrigation system (NRCS 2013).

16.2.5.1 Components of Drip Irrigation Unit

A drip irrigation unit usually consists of the following 3 main components:

1. Control head
2. Pipe network
3. Emitters (Arora 2014; Kumar and Vishal Kumar 2012).

16.3   Methodology

16.3.1   Technique for Order of Preference by Similarity to Ideal Solution (TOPSIS)

This study will use a fuzzy TOPSIS technique to analyse different irrigation water application methods or systems. In the TOPSIS method, we usually assume some rating and weights which are normally represented numerically for any problem solution to be made by a single decision-maker. But for a multiple decision making, complexity do arise due to the fact that preferred solution in this case need to be agreed by different interest groups or people with different goals and opinion. This is systematically described below at different steps.

16.3.1.1   Step 1: Construction of the Decision Matrix and the Determination of the Weight of Each Criteria

In step 1, a decision matrix $X = X_{ij}$ and a weighing vector $W = [w_1, w_2, \ldots w_n]$ are chosen. Where $X_{ij} \in \Re$, $W_j \in \Re$ and $w_1 + w_2 + \ldots w_n = 1$.

The criteria of the function can be either a cost function (less cost better result) or a benefit function (more criteria better results).

16.3.1.2   Step 2 Calculation of the Normalized Decision Matrix

In this step, comparison is made across all criteria to be used; this is done by transforming all attribute dimensions into non-dimensional attributes. To make all scores into normalize form, a transformation is been made for each evaluation matrix X, because the majority of the criteria are usually measured in various units. One of the several known standards formulas out of many is used for the normalisation of these values. The most common method used for this calculation is the normalized value n_{ij} and is given by;

$$n_{ij} = \frac{X_{ij}}{\sqrt{\sum_{i=1}^{m} X_{ij}^2}} \tag{16.1}$$

$$n_{ij} = \frac{X_{ij}}{\max_i X_{ij}} \tag{16.2}$$

$$n_{ij} = \begin{cases} \dfrac{x_{ij} - \min_i x_{ij}}{\max_i x_{ij} - \min_i x_{ij}} \\[2ex] \dfrac{\max_i x_{ij} - x_{ij}}{\max_i x_{ij} - \min_i x_{ij}} \end{cases} \tag{16.3}$$

if C_i is a benefit criterion and if C_i is a cost criterion
For $i = 1, \ldots, m; j = 1, \ldots, n$.

16.3.1.3   Step 3 How to Calculate the Weighted Normalized Decision Matrix

To calculate the weighted normalized value v_{ij}, this is done in the following way:

$$v_{ij} = w_j n_{ij} \tag{16.4}$$

For $i = 1, \ldots, m; j = 1, \ldots, n$.
Where w_j is the weight of the j-th criterion, $\sum\limits_{j=1}^{n} w_j = 1$.

16.3.1.4   Step 4 Determination of the Positive Ideal and Negative Ideal Solutions

In this step, both the positive and negative ideal alternatives i.e. the extreme performance on each criterion and the reverse extreme performance on each criterion were respectively identified. The ideal positive solution maximises the benefit criteria and minimizes the cost criteria, whereas the negative ideal solution maximizes the cost criteria and minimizes the benefit criteria.
For the Positive ideal solution A^+ it has the form:

$$A^+ = \left(v_1^+, v_2^+, \ldots, v_n^+\right) = \left[\left[\max_i v_{ij} | j \in I\right], \left[\min_i v_{ij} | j \in J\right]\right] \tag{16.5}$$

While the Negative ideal solution A^- has the form:

$$A^- = \left(v_1^- v_2^-, \ldots, v_n^-\right) = \left[\left[\min_i v_{ij} | j \in I\right], \left[\max_i v_{ij} | j \in J\right]\right] \tag{16.6}$$

where I and J in the above equation are associated with benefit criteria and cost criteria, respectively, $i = 1, \ldots, m; j = 1, \ldots, n$.

16.3.1.5   Step 5 How to Calculate the Separation Measures from the Positive Ideal Solution and the Negative Ideal Solution

In the TOPSIS method, to calculate the separation measures from the positive and negative ideal solution, we apply a number of distance metrics. For the separation of each alternative from the positive ideal solution the following equation is used;

$$d_i^+ = \left(\sum_{j=1}^{n} \left(v_{ij} - v_j^+ \right)^p \right)^{1/p} , \ldots i = 1, 2, \ldots, m. \qquad (16.7)$$

While for the separation of each alternative from the negative ideal solution, the following equation is used;

$$d_i^+ = \left(\sum_{j=1}^{n} \left(v_{ij} - v_j^- \right)^p \right)^{1/p} , \ldots i = 1, 2, \ldots, m. \qquad (16.8)$$

where $p \geq 1$. for $p = 2$ we have the most used traditional n-dimensional Euclidean metric.

$$d_i^+ = \sqrt{ \sum_{j=1}^{n} \left(v_{ij} - v_j^+ \right)^2 }, \ldots i = 1, 2, \ldots, m. \qquad (16.9)$$

$$d_i^+ = \sqrt{ \sum_{j=1}^{n} \left(v_{ij} - v_j^- \right)^2 }, \ldots i = 1, 2, \ldots, m. \qquad (16.10)$$

16.3.1.6   Step 6 How We Calculate the Relative Closeness to the Positive Ideal Solution

To calculate the relative closeness of the i-th alternative A_J with respect to A^+ is defined as;

$$R_i = \frac{d_i^-}{d_i^- + d_i^+} \qquad (16.11)$$

where $0 \leq R_i \leq 1, i = 1, 2, \ldots, m$.

16.3.1.7   Step 7 How We Rank the Preference Order or Select the Alternative Closest to 1

A set of alternatives now can be ranked by the descending order of the value of R_i.

To decide the ranking for each alternative, we use the relative closeness to the positive ideal solution. This consists of the maximum value of the alternative and the minimum value of the alternative where the aim of the criteria is maximization and minimization respectively. For negative ideal solution, this consists of the worst possible solution in terms of each criterion. It comprises of the maximum value of the alternatives and minimum value of the alternatives where the aim of the criteria is minimization and maximization respectively. Positive ideal solution is termed the best option, but it is very difficult in a real life problem to have any option having best values in terms of each criterion. With this, the TOPSIS model bring about a solution and regard any preferred alternative as the best when it is closer to the positive ideal solution and far away from the negative ideal solution. All this is done in terms of what is called Euclidian distance. From this we can understand that since the Relative closeness to the positive ideal solution is a rate that shows the closeness to the positive ideal solution while considering the closeness to the negative ideal solution, it gives the net ranking results. To rank a set of criteria, we need to convert some criteria as some are quantifiable while other may be qualitative. To do this, a cardinal scale of 0–5 is used (0 representing the worst and 5 representing the best) to transform all qualitative criteria to quantitative values, as shown in Table 16.1. The importance weights of the criteria has been defined based on the fuzzy scale (Nuhu 2020) corresponding to the opinion of the experts (Table 16.2).

After collecting all the quantitative and the qualitative data for different irrigation water application methods, the MCDM tool TOPSIS first normalize all the data as shown in Table 16.3.

16.4   Results and Discussion

For the interpretation of the result, better alternatives are observed with a higher value of the relative closeness. In this study, Table 16.4 shows the distance of the alternative from both the positive and negative ideal solutions. Table 16.5. provides the ranking of irrigation water application methods with basin method being the best method and most preferred alternative according to TOPSIS. Basin is having a score of 0.6097 and ranked as the best after considering all the criteria used in this study. Basin method is good and requires a level soil surface with uniform soil texture and adequate water supply. Followed by boarder method with a score of 0.5201, which is also a good irrigation method for level soils according to many literatures and can be used in both developed and developing countries as less energy is required by this methods. The least preferred method is drip which requires more cost of installation, more energy, more supervision and its requirement to clean and less turbid water compared to both the latter methods. Selecting an irrigation system to use is very difficult, as this varies with the peoples decisions.

The weight of the importance has been selected according to the expert's opinion. Weights can also be altered by decision makers according to their needs for better ranking result and on considering some other factors such as cost of land, season variations and many more.

Table 16.1 Decision matrix of the irrigation water application methods

Preference											
Min/max	Max	Min	Min	Max	Min	Max	Max	Max	Max	Max	Min
Importance weights	VH	M	VH	H	VH	VH	H	VH	M	H	H
Techniques/criteria	Water supply	Labour requirement	Slope (%)	Depth of irrigation water (mm)	Energy consumption	Water appl. efficiency (%)	Water quality	Attainable irrig efficiency	Aesthetics	Operational skill requirement	Cost (construction and maintenance)
Sprinkler	1	2	5	5	5	100	5	90	5	5	5
Drip	1	2	5	5	5	100	4	90	4	4	4
Furrow	3	5	2	35	1	90	1	72	2	1	1
Boarder	3	5	1	45	3	77	1	77	3	1	1
Basin	5	5	0.1	55	2	77	1	85	2	1	1

Table 16.2 Linguistic fuzzy scale

Linguistic scale for evaluation	Triangular fuzzy scale	
Very high [VH]	(0.75, 1, 1)	Water supply, slope, energy consumption, water application efficiency, attainable irrigation efficiency.
High [H]	(0.50, 0.75, 1)	Depth of irrigation water, water quality, operational skill requirement, cost(constr and maintenance)
Medium [M]	(0.25, 0.50, 0.75)	Aesthetics, labour requirement
Low [L]	(0, 0.25, 0.50)	
Very low [VL]	(0, 0, 0.25)	

16.5   Conclusions

For an improved water resource management in water irrigation application all over the world, a decision making support system is needed. Decision making needs reliable ideas than random guesses, as mere guesses and assumptions can leads to problem creation and system failures. In this research, five different irrigation water application methods were ranked using TOPSIS MCDM tool. The system have been proven to be one of the best effective instruments that can be used when selecting the most appropriate among many according to research. For this reason, this study has presented a framework using the TOPSIS algorithm as an effective supporting tool for decision making. From the results obtained, basin method has been the most preferred method followed by border method for all the irrigation water application methods. Comparing the TOPSIS MCDM tool with other decision making tools such as Analytic Hierarchy Process (AHP) and Fuzzy PROMETHEE will be of great importance and will serve as a check for making proper decision. Decision-makers, governmental, non-governmental organisation and other water users can use this decision making algorithm in making decisions in water management and will help them to keep face with other competitors not only in the engineering fields but also in the economy.

Table 16.3 Normalized weighted scores table for the alternatives for irrigation water application methods

Technique/criteria	Water supply	Labour requirement	Slope/topography (%)	Depth of irrigation water (mm)	Energy consumption	Water appl. Efficiency (%)	Water Quality	Attainable irrig. efficiency	Aesthetics	Operational skill Requirement	Cost (construction and maintenance)
Sprinkler	0.0159	0.0128	0.0721	0.0055	0.0669	0.0535	0.0657	0.0518	0.0382	0.0657	0.0657
Drip	0.0159	0.0128	0.0721	0.0055	0.0669	0.0535	0.0526	0.0518	0.0305	0.0526	0.0526
Furrow	0.0478	0.0319	0.0288	0.0384	0.0134	0.0482	0.0131	0.0414	0.0153	0.0131	0.0131
Boarder	0.0478	0.0319	0.0144	0.0493	0.0401	0.0412	0.0131	0.0443	0.0229	0.0131	0.0131
Basin	0.0797	0.0319	0.0014	0.0603	0.0267	0.0412	0.0131	0.0489	0.0153	0.0131	0.0131

Table 16.4 Distance of alternative methods from the positive ideal solution (d_i^+) and the negative ideal solution (d_i^-)

	d_i^+	d_i^-	$d_i^+ + d_i^-$
Sprinkler	0.1330	0.0817	0.2147
Drip	0.1300	0.0644	0.1944
Furrow	0.0938	0.0982	0.1920
Boarder	0.0914	0.0991	0.1905
Basin	0.0822	0.1285	0.2107

Table 16.5 Relative closeness to the positive ideal solution for each of the alternative methods

Ranking	Methods	R_i
1	Basin	0.6097
2	Boarder	0.5201
3	Furrow	0.5114
4	Sprinkler	0.3806
5	Drip	0.3312

References

Arora VK (2014) Managing water crisis for sustainable crop productivity in Punjab: an overview. J Res (Punjab Agric. University) 45:17–21. (November). https://doi.org/10.13140/2.1.4097.6001

Bjorneberg DL (2013) Author' s personal copy IRRIGATIONl methods ☆, reference module in earth systems and environmental sciences. Elsevier Inc. https://doi.org/10.1016/b978-0-12-409 548-9.05195-2

Khanna M, Malano HM (2006) Modelling of basin irrigation systems : a review modelling of basin irrigation systems: a review , (May). https://doi.org/10.1016/j.agwat.2005.10.003

Kumar S, Vishal Kumar RKS (2012) Fundamentals of agricultural engineering. Kalyani 2012

Miao Q, Shi H, Gonçalves JM (2018) Basin irrigation design with multi-criteria analysis focusing on water saving and economic returns : application to wheat in Hetao. Yellow River Basin. https://doi.org/10.3390/w10010067

NRCS (2012) Surface IRRIGATION. In: National engineering handbook (Chap 4), Washington, DC, USDA

NRCS (2013) Microirrigation. In: Part 623 Irrigation national engineering handbook (Chap 7), Washington, DC USDA

NRCS (2016) Sprinkler irrigation. In: Nation engineering handbook (Chap 11), Washington, DC; USDA

Nuhu M et al (2020) Evaluation of wastewater treatment technologies using TOPSIS. 177(May 2019), 416–422. https://doi.org/10.5004/dwt.2020.25172

Phocaides A (2007) Handbook on pressurized irrigation techniques. In: Food and agriculture organisation of the United Nations

Zerihun D, Sanchez CA, Yitayew M (2013) Analysis and design of border irrigation systems' (May 2014). https://doi.org/10.13031/2013.20009

Chapter 17
Comparative Analysis of Flexible Pavement Design Methods Using Fuzzy PROMETHEE

Ibrahim Khalil Umar, Hüseyin Gökçekuş, and Dilber Uzun Ozsahiņ

Abstract Various methods for flexible pavement design were developed over the years ranging from empirical, mechanistic and the recently the mechanistic-empirical method. In this paper seven different pavement design methods were studied and each of the methods was used to determine the thickness of a flexible pavement. Fuzzy PROMETHEE decision technique was used to identify the most suitable method for designing the pavement. The criteria employed in the selection of the best method for the pavement design includes pavement thickness obtained from the design result, environmental consideration and pavement performance in the design procedure, recommended design life, minimum strength requirement for subbase and base material. Road Note 29 (RN29) was found to be the most preferred method with HD26/01 been the second. HD26/01 will be recommended for design of new road, since research proved total failure of the road pavements designed with RN29 after 20 years.

Keywords Subgrade · Flexible pavement · Axle load · Fuzzy PROMETHEE

I. K. Umar (✉) · H. Gökçekuş
Faculty of Civil and Environmental Engineering, Near East University, Nicosia, Turkish Republic of Northern Cyprus, Turkey
e-mail: 20178443@std.neu.edu.tr

H. Gökçekuş
e-mail: huseyin.gokcekus@neu.edu.tr

D. Uzun Ozsahiņ
DESAM Institute, Near East University, Nicosia, Turkish Republic of Northern Cyprus, Turkey

Department of Biomedical Engineering, Near East University, Nicosia, Turkish Republic of Northern Cyprus, Turkey

Medical Diagnostic Imaging Department, College of Health Science, University of Sharjah, Sharjah, United Arab Emirates

D. Uzun Ozsahiņ
e-mail: dilber.uzunozsahin@neu.edu.tr

© The Author(s), under exclusive license to Springer Nature Switzerland AG 2021
D. Uzun Ozsahin et al. (eds.), *Application of Multi-Criteria Decision Analysis in Environmental and Civil Engineering*, Professional Practice in Earth Sciences,
https://doi.org/10.1007/978-3-030-64765-0_17

17.1 Introduction

Flexible pavement is a structure designed for a certain period known as design life to resist traffic and environment. The structure is designed to safeguard the subgrade and maintain safety and cost of operation within a reasonable limit. The fundamental inputs for designing the pavement include the strength of the subgrade, environmental conditions, strength of the layer materials and the predicted traffic load (McElvaney and Snaith 2012). The pavements could be designed by many different empirical procedures, even though they are convenient to use, research proved that, some of these methods provide undesirable results. This major drawback of the procedures has led to the development of mechanistic procedures over the last 40 years (Allen et al. 2015).

Many comparative studies were conducted to compare the different flexible pavement design procedures. For example, Yahya et al. compares cost of flexible pavement in Nigeria designed using the contemporary methods and found CBR method to be cost effective (Yahaya et al. 2018). Saha et al. 2012 compares the American association of state highway and transportation officials (AASHTO) mechanistic empirical design guide (MEPDG) and the Alberta transportation flexible pavement design procedure. It was found that, when using the MEPDG, only the cases with a strong subgrade material and a low level of traffic meet the default limit value for total pavement rutting. El-Badawy et al. (2011) also compares Idaho Pavement Design Procedure with AASHTO 1993 and MEPDG Methods. The findings shows that, the Idaho pavement design method overestimates the pavement thickness in comparison to the AASHTO 1993 and MEPDG. Chidozie and Joshua (2016) compared the traffic loading obtained using the Road Note 29, HD 26/01, and LR1132 procedures for designing flexible pavement. Perraton et al. (2010) made a comparison between pavement design methods from a fatigue point of view.

In this paper a comparative analysis of some flexible pavement design methods was done using fuzzy preference ranking organization method for enrichment evaluations (PROMETHEE) technique to identify the best method to be used for design flexible pavement considering multiple criteria.

17.2 Flexible Pavement Design Method

17.2.1 Asphalt Institute Method

Multi-layered elastic system is used to represent the pavement structure in the Asphalt Institute design method. The load from the traffic is applied as a uniform vertical stress through the tire which is spread by the different pavement layers and finally to the subgrade as a lower stress. Established theories, experience and test results are applied to determine the two stress–strain situations. Firstly, the change of the stress through the pavement layer and secondly the tensile and compressive stresses and strains imposed on the asphalt due to the deflection caused by wheel loads. The pavement design charts were developed based on the maximum strain at the bottom of the surface layer and the highest vertical compressive strains at the top of the subgrade layer. The basic principle in the asphalt institute method is determining the least thickness of the pavement surface to sufficiently resist the compressive stresses at the surface and subgrade layer, and the tensile strain below the asphalt layer developed. Design charts for different range of traffic loads have been prepared (Nicholas and Lester 2002). The four different criteria required to determine proper thickness of the pavement layer in Asphalt Institute method are traffic loading of the pavement in ESALs; material properties, especially the resilient modulus of the soil at the foundation layer; the mean annual air temperature of the pavement location, the desired base materials: granular or hot mix asphalt (Lavin 2003). The detail procedure for determining pavement thickness using the asphalt institute method can be found in (Nicholas and Lester 2002).

17.2.2 AASHTO 1993 Method

This is an empirical method for pavement design developed by AASHTO. The result of the field performance of roads conducted in Ottawa and Illinois in 1958–60 forms the basis of this method. The approach of this method is to design for specific loss in serviceability at the end of the pavement intended life (AASHTO 2001). The stresses due to variations in temperature and moisture, traffic loading, time constraints and other design variables are used to design the pavement layers as to make the pavement maintain the required serviceability throughout the pavement design life (Khan et al. 2012). Inputs for the design of flexible pavement in AASHTO procedure were classified into (i) design variables (time constraints, traffic, reliability, environmental impacts) (ii) performance criteria (serviceability, allowable rutting, aggregate loss) (iii) material properties for structural design (effective roadbed soil resilient modulus, effective modulus of subgrade reaction, pavement layer material characterization, PCC modulus of rupture, layer coefficient) and (iv) the structural characteristics (drainage, load transfer, loss of support).

17.2.3 California Bearing Ratio CBR Method

In 1928, the highway division in California developed the first empirical method to design flexible pavement which is known as the CBR method. United States Army Corps adopted the method in 1945. The thickness of the pavement constituents is assessed by the strength of the subgrade expressed as CBR index value (Pereira and Pais 2017). The ratio of test load to the standard load for specified plunger penetration expressed in percentage is defined as CBR (Gill and Maharaj 2015). Thickness of the surface, base and subbase layers are obtained using results of the CBR test conducted on the subgrade, base and subbase materials combined with the empirical charts. Pavement thickness depends on the CBR of value of the subgrade and the constituent materials. The higher the CBR value the thinner the layer thickness and vice versa (Kumar and Pavithra 2016). The procedure for determining pavement thickness in this method involved determination of Base Year Traffic Counts, Design Life and Traffic Growth Rates, CBR value of subgrade, base and subbase materials, number of commercial vehicles per day (Yahaya et al. 2018). Equation 1 gives the relationship between layer thickness and CBR values (Ekwulo and Eme 2009).

$$t = \sqrt{w\left[\frac{1}{8.1CBR} - \frac{1}{p\pi}\right]} \tag{17.1}$$

17.2.4 Group Index (GI) Method

High Research Board suggests the Group Index method for estimating the thickness of flexible pavement. Pavement thickness is obtained using group index value and the vehicular load expressed in terms of commercial vehicles (Khan et al. 2012). The lower the GI of the subgrade, the higher the strength and the lower the subbase thickness and vice versa (Gill and Maharaj 2015). Equation 2 gives the expression for determining the GI of the soil.

$$GI = 0.2a + 0.005ac + 0.01bd \tag{17.2}$$

where, a = part of the material passing through 0.074 mm sieve that is more than 35% and not more than 75%, b = proportion of material passing through 0.074 mm sieve that is higher than 15 and not more than 35%, c = value of liquid limit between 40 and 60, d = value of plasticity index exceeding 10 and less than 30. The design traffic load is obtained using Eq. 3 for the design life of the pavement.

$$A = PX(1 + r)^n \tag{17.3}$$

where, A = Design traffic load in vehicles/day for the design period, P = number of commercial vehicles in day expressed in vehicles/day, n = Design life in years and r = Annual growth rate (%).

17.2.5 California (Hveem) Design Method

The method was developed in 1940s based on data gathered from test road pavement, experience and theory. It is used in western states. It was modified many times to incorporate changes in the traffic characteristics. Initially, the design was to avoid distortion and plastic deformation of the pavement surface but the modified methodology includes reduction of early fatigue cracks to the minimum. The factors considered are (i) Strength of construction materials (ii) Strength of subgrade material (iii) Traffic load.

The traffic load S Initially calculated as the ESAL-that is, the total number of 18,000-lb axle loads in one direction-as discussed earlier, and then converted to a traffic index (TI), where

$$TI = 9.0 \times \left(\frac{ESAL}{10^6}\right)^{0.119} \qquad (17.4)$$

The design objective is to determine the total thickness of material required above the subgrade to carry the projected traffic load. This thickness is determined in terms of gravel equivalent (GE) in feet, which is given as

$$GE = 0.0032(TI)(100 - R) \qquad (17.5)$$

where TI = traffic index, GE = material thickness needed on a given layer in terms of gravel equivalent (feet), R = resistance value of the supporting layer material, mostly determined at a discharge pressure of 300 Ib/in^2. Thickness of each of the layer is obtained by dividing GE for that layer by the GE factor G_f for the material used in the layer. A check must be done to ensure that the requirements for expansion pressure is adequate for the layer thickness obtained.

17.2.6 Hd26/01

The United Kingdom department of transport developed this standard procedure in 2001. It is the modification of the LR 1132 based on availability of recent construction materials and innovative research. The most important and primary factor of this method of design is the traffic assessment (Chidozie and Joshua 2016). The weighted yearly traffic is computed using Eq. 6 and the corresponding design traffic is obtained using Eq. 17.7.

$$T_i = 365 \times Y \times F \times W \times G \times P \times 10^{-6} \, \text{msa} \qquad (17.6)$$

$$\text{Design Traffic } (T) = \sum T_i \qquad (17.7)$$

where: Y = Design Period in Years, F = Traffic flow for each class of traffic at opening, W = Wear Factor for each traffic class (W_N in case of new design or W_M in case of Maintenance), G = Growth Factor, P = Percentage of vehicles in the heaviest loaded lane (Highways Agency 2006).

17.2.7   Road Note 29

The officially recognized design procedure for the design of flexible pavement throughout the 1970 and up to early 1980s is the Road Note 29. In 1973, the department of Environment of the United Kingdom published the RN29 for first time. The procedure takes into consideration increase in traffic volume and axle loads, strength of the subgrade material, while differentiating performance of various materials used for road bases. In this method it is assumed that, the road base satisfies the overall strength required for the whole pavement structure while the surfacing is just to provide regularity and surface texture without adding any significant strength to the pavement structure (Rogers 2003).

17.3   Methodology

17.3.1   Design of Flexible Pavement

The analysis traffic data was obtained from the research conducted by Yahya et al. in Nigeria where an AADT of 4929 passenger car per day was obtained. The design was based on a subgrade with a CBR value of 13%. The traffic growth and design life were 7.5% and 20 years respectively (Yahaya et al. 2018). A subbase and base materials with CBR value of 25% and 100% were used where applicable for all methods. The total thickness obtained from each method using appropriate charts and corresponding axle loads were presented in Table 17.3 as part of the selection criteria.

17.3.2   Fuzzy PROMETHEE (F-PROMETHEE)

Brans et al. developed PROMETHEE procedure for making decision where various alternatives exist. The procedure was based on comparing pair of the alternatives with respect to the selected criteria. The model is among the less complicated in terms of application and conception among many multi criteria decision making tools (Uzun et al. 2018). The F- PROMETHEE combines both PROMETHEE and fuzzy logic. This make its application possible in many decision-making scenarios. The method was recommended for comparing non-numeric alternatives by Wang, et al. (2008). Collecting satisfactory data to fully examine a problem and make appropriate decision is sometimes difficult in actual life. But with fuzzy sets, the decision maker can be able to examine the system in a fuzzy condition which is practical. Ozsahin et al. 2017 gives detail discussion and examples of practical application of the F-PROMETHEE method used in this study.

Table 17.1 Linguistic fuzzy scale

Linguistic scale for evaluation	Triangular fuzzy scale	Priority ratings of criteria
Very high (VH)	0.75, 1, 1	Thickness, environmental, base CBR
Important (H)	0.5, 0.75, 1	Pavement performance
Medium (M)	0.25, 0.50, 0.75	High traffic, subbase CBR,
Low (L)	0, 0.25, 0.5	Design life
Very low (VL)	0, 0, 0.25	

Fuzzy scale in Table 17.1 was used to compare the defined criteria of the pavement methods effectively in order to get significance of each criteria. Yager index was employed to defuzzify the triangular fuzzy numbers to forecast the weight of each criterion.

After gathering the parameters for the comparison of the flexible pavement design methods, Gaussian preference function was utilized for each criterion as presented in Table 17.2. Visual PROMETHEE decision lab program was then applied.

17.4 Result and Discussion

Table 17.3 and Fig. 17.1 provides the ranking of the flexible pavement design method with RN 29 been the best method to be used considering all the criteria followed by the HD 26/01 design method which has its origin from RN29 to LR1132, and finally HD 26/01 (Rogers 2003). The higher thickness requirement in HD26/01 is the major reason why it becomes the second-best method, since cost is directly related to the overall pavement thickness. The Group index method becomes the least preferred method, this is mainly due to the fact that physical properties of the material are considered rather than strength.

17.5 Conclusion

The fuzzy PROMETHEE decision making technique has proved to be effective in selecting most preferred methods for pavement design. Road Note 29 was found to the best method with HD26/01 been the second most preferred method for pavement design despite CBR method giving the least pavement thickness. This research will be recommended for design of new roads, since research proved total failure for roads designed with RN29 after 20 years.

Table 17.2 Visual PROMETHEE for selection of best pavement design method

Design method	Total thickness (mm)	Design life (years)	Minimum subbase CBR (%)	Minimum base CBR (%)	Recommended for traffic	Pavement performance	Environmental consideration
Preference							
Min/Max	Min	Max	Max	Max	Max	Max	Max
Weight	*0.92*	*0.25*	*0.50*	*0.92*	*0.50*	*0.75*	*0.92*
Asphalt Institute	500	20	20	80	High	Yes	Yes
AASHTO	560	20	30	80	High	Yes	Yes
CBR	300	10	25	80	Low	No	No
Group index	400	10	30	80	Low	No	No
Hveem	860	20	30	80	Medium	Yes	No
Road note 29	400	20	30	100	Medium	No	No
HD 26/01	760	40	30	100	High	No	Yes

Table 17.3 Complete ranking of flexible pavement design methods

Ranking	Methods	Net flow	Positive outflow ranking	Negative outflow ranking
1	Road note 29	0.3059	0.3583	0.0523
2	HD 26/01	0.1267	0.3008	0.1742
3	CBR	0.0984	0.2685	0.1701
4	AASHTO	−0.0361	0.1748	0.2109
5	Asphalt Institute	−0.0851	0.1764	0.2616
6	Hveem method	−0.1673	0.1063	0.2736
7	Group index	−0.2424	0.1412	0.3836

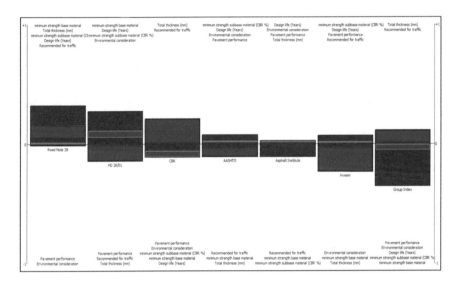

Fig. 17.1 Evaluation of flexible pavement design methods

References

AASHTO (2001) AASHTO guide for design of pavement structures. AASHTO, Washington, DC

Allen DH, Little DN, Soares RF, Berthelot C (2015) Multi-scale computational model for design of flexible pavement–part I: expanding multi-scaling. Int J Pavement Eng 18:1–12

Chidozie NM, Joshua CC (2016) Traffic assessment for pavement structural design using RN29, LR1132, and HD26/01 methods. 1:47–52. https://doi.org/10.11648/j.ajtte.20160104.12

Ekwulo EO, Eme DB (2009) Fatigue and rutting strain analysis of flexible pavements designed using CBR methods. Afr J Environ Sci Technol 3:412–421. https://doi.org/10.1080/14680629.2007.9690094

El-Badawy S, Bayomy MF, Santi M, Clawson CW (2011) Comparison of idaho pavement design procedure with AASHTO 1993 and MEPDG methods. In: ASCE Proceedings of T & DI Congress. Chicago, Illinois, USA

Gill S, Maharaj DK (2015) Comparative study of design charts for flexible pavement. Int Res J Eng Technol 2:339–348

Highways Agency (2006) Pavement design and maintenance Part 3 HD 26/06. In: Design manual for road and bridges. Department of Transport, Middlesex

Khan RU, Khan MI, Khan AU (2012) Software development (PAKPAVE) for flexible pavement design, pp 1–5

Kumar RV, Pavithra M (2016) Experimental study on design of flexible pavement using CBR method. 6890:63–68

Lavin PG (2003) Asphalt pavement: a practical guide to design, production, and maintenance for engineers and architects. Spon Press

McElvaney J, Snaith MS (2012) Analytical design of flexible pavements. In: Highways, pp 395–423

Nicholas JG, Lester AH (2002) Traffic and highway engineering. Third Edit, Brools/Cole, Virginia

Ozsahin DU, Uzun B, Musa MS et al (2017) Evaluating nuclear medicine imaging devices using fuzzy PROMETHEE method. Proc Comput Sci 120:699–705. https://doi.org/10.1016/j.procs.2017.11.298

Pereira P, Pais J (2017) Main flexible pavement and mix design methods in Europe and challenges for the development of an European method. J Traffic Transp Eng 4:316–346. https://doi.org/10.1016/j.jtte.2017.06.001

Perraton D, Baaj H, Carter A (2010) Comparison of some pavement design mathods from a fatigue point of view: effct of fatigue properties of asphalt materials. Road Mater Pavement Des 11:833–861. https://doi.org/10.3166/rmpd.11.833-861

Rogers M (2003) Highway engineering. Blackwell Science Oxford, UK

Saha J, Nassiri S, Soleymani H, Bayat A (2012) A comparative study between the alberta transportation flexible pavement design and the MEPDG. Int J Pavement Res Technol 5:379–385

Uzun D, Uzun B, Sani M, Ozsahin I (2018) Evaluating x-ray based medical imaging devices with fuzzy preference ranking organization method for enrichment evaluations. Int J Adv Comput Sci Appl 9. https://doi.org/10.14569/IJACSA.2018.090302

Wang TC, Chen LY, Chen YH (2008) Applying fuzzy PROMETHEE method for evaluating is outsourcing suppliers. In: Proceedings—5th international conference on fuzzy systems and knowledge discovery, FSKD 2008, pp 361–365

Yahaya AB, Ako T, Jimoh YA (2018) A comparative study of contemporary flexible pavement design methods in nigeria based on costs. Int J Civ Eng Constr Sci 5:7–16

Chapter 18
Using Clustering of Panel Data to Examine Housing Demand of Expatriate Turks and Foreigners: An Application of k-prototype Algorithm

Özlem Akay, Cahit Çelik, and Gülsen Kıral

Abstract In recent years, development in the Turkey's economic structure and implementation of the "economic and financial policy" shows significant effects on the housing sector. For this reason, according to economic factors affecting housing demand, it is important to find cities and countries showing common characteristics and to examine the developments in the housing market. In this study, it was aimed to determine the common characteristics of the provinces that preferred by the Turks living abroad and foreigners and of the countries that prefer housing demand from these provinces in Turkey. Two different panel data sets were created for expatriate Turks and foreigners who prefer the demand for housing in Turkey. Cluster analysis was performed to these data sets using k-prototypes algorithm Cluster validity indexes were calculated to determine the appropriate number of clusters for both data sets. According to the findings, while the optimal number of clusters for data set defined for the expatriate Turk preferring housing demand in Turkey was found to be six, the optimal number of clusters for the data set defined for foreign countries preferring housing demand from Turkey was found to be nine.

Keywords Housing demand · Clustering of panel data · k-prototypes algorithm · Turkey

Ö. Akay (✉)
Department of Statistics, Cukurova University, 01330 Adana, Turkey
e-mail: oakay@cu.edu.tr

C. Çelik · G. Kıral
Department of Econometrics, Cukurova University, 01330 Adana, Turkey
e-mail: cahit.celik@outlook.com

G. Kıral
e-mail: gkiral@cu.edu.tr

18.1 Introduction

The sensitivity of the construction and housing sector to the general economic conditions is different in each country. The construction and housing sector in the global market has extensive experience and potential. The construction sector is the locomotive sector since it affects more than 250 sub-sectors connected to it. The construction sector, which is based on domestic capital, is also absorbing unemployment due to create employment related to hundreds of occupations. The employment potential created for production by Turkish construction and sub-sector components provides important advantages for the national economy. In addition, housing not only contains economic and demographic factors, but also has socioeconomic, sociodemographic and sociopsychological qualities. According to TurkStat data, the expatriate Turks, which turned the rise of foreign exchange into an opportunity to invest in real estate in their countries, accelerated the purchase of immovable properties, especially in Izmir and Istanbul. Turkish real estate purchases increased by 23% compared to the same period of 2016. After the amendment of the reciprocity law, there was an explosion in selling property to foreigners.

Middle East countries and other countries buying heavily housing from Turkey prefer green and cool regions such as Izmit and Trabzon as well as Istanbul, Antalya, Izmir and Yalova. According to TurkStat data, existing home sales to foreign countries from Turkey, the record levels reached 40 thousand in 2018. This increase in housing sales was influenced by the increase in exchange rates of 2018, a value-added tax (VAT) exemption for non-Turkish citizens, the reduction of the value to be paid for foreigners to obtain Turkish citizenship from 1 million dollars to 250 thousand dollars, and international promotion and fair activities. Turkish citizens living in Germany and foreigners wishing to acquire immovable property in Turkey is able to land transactions with Berlin Land Registry and Cadastre Office established in Consulate General of Berlin by General Directorate of Land Registry and Cadastre. In addition to procurement transactions, deed registration sample, mortgage, donation transactions and title deed registration, exchange, transfer, applications and map related to cadastral data can be done in this agency. The land registry and cadastre offices planned to be installed in any two countries, Denmark, Austria, Germany, Belgium, Greece, Russia, Britain, the Netherlands, Norway, Qatar and Turkey. While determining these countries, the number of Turkish citizens living in the country, the number of foreigners who acquire real estate, high potential countries and economic and historical relations are taken into consideration.

In this study, panel data sets were created by obtaining economic indexes data of the expatriate Turks demanding house in Turkey for monthly in 2013–2016 and of the foreign countries demand for housing in Turkey for monthly in 2015–2016. The data used in the analysis were taken from TURKSTAT (Turkey Statistical Institute). When households living outside Turkey are demanding house in Turkey, according to their expectation, cluster analysis was performed using the k-prototype algorithm to find the similar characteristics of preferred provinces and the countries they live in. R program was used for analysis.

The study consists of five section. In the first section of the study, the importance of housing demand in terms of Turkey's economy, Turks living abroad and foreign countries interest in housing demand from Turkey are described. The second part consists of extensive literature research. In the third part of the study, clustering of panel data, k-prototype algorithm and cluster validity indexes are disclosed to be used in the application phase. The fourth part of the research consists of the application Clustering of panel data were made by using k-prototype algorithm for both expatriate Turks and foreign countries with the factors affecting housing demand and interpreted in detail. In the last section of the study, the results are given. This study carries original research quality since clustering of panel data were made by using k-prototypes algorithm in order to determine interest of the individuals living outside Turkey who are demanding house in Turkey.

18.2   Literature Review

Carliner (1973) offered new evidence based on better data than has been available to earlier researchers. Using a four-year panel study which followed up movers, permanent income was defined and calculated in two ways. Then regressions were run on house value and rent on permanent income, price, age, race and sex of head. The results obtained are robust with respect to the definition of permanent income, and considerably lower than results from time series analysis or from cross-section studies that relied on grouped rather than individual data.

Hanushek and Quigley (1980) focused on household price sensitivity. However other changes such as income or family size in household conditions clearly affect housing consumption. Actually, given limited longitudinal data, information about other demand adjustments provides valuable information about consumption dynamics, given limited longitudinal data. Price changes can be seen as one of the various external effects on housing demand.

Goodman (1988) addressesed the determination of permanent income, housing price,housing demand and tenure choice. Full housing demand elasticities incorporate the interactive effects among the four stages of the model. Price and income have major effects in the tenure choice equation. Socio demographic variables, such asage, have complex effects that may be lost in simpler forms of estimation.

Bajari and Kahn (2005) presented a three-stage, nonparametric estimation procedure to recover willingness to pay for housing attributes. Firstly, they estimated a nonparametric hedonic home price function. Secondly, they recovered each consumer's taste parameters for product characteristics using first-order conditions for utility maximization. Finally, they estimated the distribution of household tastes as a function of household demographics.

Nunes and Serrasqueiro (2007) showed that internal and external financing are not perfect substitutes using panel data for the period 1999–2003, not corroborating the theorem of Modigliani and Miller.

Nunes et al. (2009) studied the profitability determinants of Portuguese service industries based on various panel models.

Davis et al. (2017) stated that households holding an FHA mortgage increased the value of the housing they purchased by approximately 2.5% using the financing of Fannie Mae and Freddie Mac. After converting the premium discount to an equivalent decrease in the mortgage rate, conditional on purchasing a home with roughly 3.4 estimates means semi-elasticity of the value of the housing purchased on the mortgage rate.

Nguyen and Nordman (2018) shaded light on the links between households' and entrepreneurs' social networks and business performance by using a unique panel of household businesses for Vietnam.

Do et al. (2019) examined the determinants of livestock assets with panel data from Vietnam. They suggested that authorising rural households to better cope with shocks contributes to reducing rural poverty and to developing livestock. Ahmad et al. (2018) investigated the determinants of housing demand in urban areas of Pakistan. Empirical analysis was performed using the 2004–05 and 2010–11 Pakistan Social and Standard of Living Measurement (PSLM) survey. The hedonic price model was used to estimate housing prices. Heckman's two-stage selection procedure is used to control the bias of selectivity between duty term selection and the amount of housing services requested. Empirical analysis shows that house prices and revenues (temporary and permanent) play an important role in determining the demand for housing units.

Huarng et al. (2019) analyzed housing demand by using Google Trends' big data as a proxy. They use to estimate a qualitative method (fuzzy set/ Qualitative Comparative Analysis, fsQCA) instead of a quantitative method. According to empirical results, although the size of the data set is small, fsQCA successfully predicts seasonal time series.

Zheng et al. (2018) estimated the income elasticity of demand for private rental housing using micro data between 1996 and 2011 from four waves of four Hong Kong census data. In order to isolate permanent and temporary income at the household level they adopted a permanent income model. They used the Heckman two-step procedure to correct selection bias and used the quantile regression (QR) approach to investigate the heterogeneity of demand elasticities between different levels of housing expenditure. Empirical results show that permanent income elasticities fall within the range of 0.536–0.698 and that temporary income shock has a positive and significant impact on rental housing demand.

Liu (2019) examined the theoretical relationship between income and home prices by using the user cost equilibrium condition. Empirically, the short-term and long-term dynamics of this relationship studied from 1991 to 2015 for 25 years in the state of New South Wales, Australia, using data for 144 Local Government Areas (LGAs). He estimated to be 1.07 the income elasticity of housing prices for the government with multi-factor panel data models and cointegration analysis.

Çelik and Kiral (2018a) applied balanced and unbalanced panel data analysis and clustering analysis methods to factors that affect housing sales in provinces of Turkey and examined significant variables by hierarchical clustering method. They also supported the study with SWOT analysis. Çelik and Kıral (2018b) used clustering of panel data and SWOT analysis to examined the socio-economic factors affecting housing sales of the provinces of Turkey in the 2008–2015 process. They determined factors affecting housing sales in provinces that exhibit similar characteristics in housing demand. The results obtained from the study showed that urbanization rate, the ratio of deposit interest rate, average household income, number of household's automobiles, stock market Istanbul 100 index, housing loan interest rateangross return rate of housing, were significant in for housing demand. Akay and Yüksel (2018) presented that the mixed panel dataset is clustered by agglomerative hierarchical algorithms based on Gower's distance and by k-prototypes . Akay and Yüksel (2019) suggested a new distance for clustering of the mixed variable panel data set containing invariant time binary variable, without performing variable conversion to avoid information loss.

18.3   Cluster Analysis of Panel Data with k-Prototype Algorithm

Panel data refer to two-dimensional data which are obtained in time series and cross section at the same time, and that means taking multiple cross sections on time series, and selecting the sample observations on cross sections at the same time (Hou and Ai 2015). The poobility of the different topics in the data is one of the important issues in the panel data. If the parameters in the regression can be considered homogeneous between different subjects, different subjects can be brought together. However, the normal situation is that subjects cannot be pooled due to high heterogeneity. Some recent studies investigated the "partial poolability" by clustering subjects into different groups so that subjects in the same cluster have homogeneous parameters (Lu and Huang 2011).

Bonzo and Hermosilla (2002) applied probability link function to advance the algorithm of the cluster, thus the cluster analysis could be effectively applied to the analysis panel data. In this study, we choose k prototype algorithm to explain the cluster analysis process of multivariable panel data.This algorithm was proposed by Huang (1998). It is straightforward to integrate the k-modes and k-means algorithms into the k-prototypes algorithm used to cluster the mixed-type objects. Since frequently encountered objects in real world databases are mixed-type objects, the k-prototypes algorithm is practically more useful. The cost function is used in conjunction with a partitioned clustering algorithm. The cost function handles mixed datasets and computes the distance between a data point and a centre of cluster in terms of two distance values—one for the numeric attributes and the other for the categorical attributes.The objective of k-prototype is to group the dataset X into k clusters by minimizing the cost function,

$$E = \sum_{l=1}^{k} \sum_{i=1}^{n} y_{il} d(X_i, Q_l) \tag{18.1}$$

Here, $Q_l = [q_{l1}, q_{l2}, \ldots, q_{lm}]$ is the representative vector or prototype for cluster l, and y_{il} is an element of a partition matrix Y_{nxl}. $d(X_i, Q_l)$ is the dissimilarity measure defined as follows:

$$d(X_i, Q_l) = \sum_{j=1}^{p} \sum_{t=1}^{T} \left(X_{ij}^r(t) - q_{lj}^r \right)^2 + \mu_l \sum_{j=p+1}^{m} \delta \left(X_{ij}^c(t) - q_{lj}^c \right) \tag{18.2}$$

where $\delta(p, q) = 0$ for $p = q$, and $\delta(p, q) = 1$ for $p \neq q$; $X_{ij}^r(t) \left(X_{ij}^c(t) \right)$ is the value of the jth numeric (categorical) attribute at the time t for the data object i; $q_{lj}^r \left(q_{lj}^c \right)$ is the prototype of the jth numeric (categorical) attribute in the cluster l;μ_l is a weight for categorical attributes in the cluster l (Ji et al. 2012). The process of the k-prototype algorithm is described as follows:

The process of the K-prototype algorithm is defined as follows:

Step 1. Randomly select k data objects from the dataset X as the initial prototype of the sets.

Step 2. For each data object in X, assign it to the cluster whose prototype is closest to that data object in terms of Eq. (18.2). After each assignment, update the prototype of the cluster.

Step 3. After all data objects have been assigned to a cluster, recalculate the similarity of the data objects with the existing prototypes. If a data object is found to belong to another cluster rather than the closest prototype, reassign that data object to that cluster and update the prototypes of both clusters.

Step 4. After the full circle test of X, terminate the algorithm if no data object has changed the sets, or else repeat step 3 (Ji et al. 2013).

Different clustering algorithms often lead to different clusters of data, even for the same algorithm, the choice of different parameters or the order of presentation of data objects can greatly affect the final clusters. Therefore, effective assessment standards and criteria are critical to reassuring users of cluster results. For all that, these evaluations provide meaningful information on how many clusters are hidden in the data. Actually, the user is faced with the dilemma of selecting the number of clusters or partitions in the underlying data. Therefore, numerous indices have been proposed to determine the number of clusters in a data set (Charrad et al. 2012).

Some clustering validity indices are used to select the optimal number of clusters. These indices are The C-Index, Dunn index, Gamma index, Gplus index, McClain index, Ptbiserial index, Silhouette index and Tau index. The minimum values of the C-Index, Gplus and McClain index are used to indicate the optimal number of clusters. The maximum values of the Dunn, Gamma, Ptbiserial, Silhouette and Tau index are used to indicate the optimal number of clusters.

18.4 Application

This study aims to determine provinces in Turkey and countries showing common feature according to the expectations the Turks living abroad and foreign countries in housing demand. For this purpose, panel data set was formed, firstly, by monthly data for the years 2013–2016 for provinces where expatriates demand housing from Turkey (Antalya, Istanbul, Trabzon, Aydin, Bursa, Other Provinces, Mersin, Mugla, Yalova, Ankara, Izmir, Sakarya) and secondly by considering economic factor indices determining the demand for housing.

Economic factor indexes, European Currency, Bullion Gold, General Import, Economic Confidence Index, Consumer Confidence Index, Real Sector Confidence Index, Construction Sector Confidence Index, Employment Index, Employees in a Paid Job are determined as the shadow variable taking a value of 1 for investment purpose provinces and 0 for the others as described in detail in Table 18.1.

Subsequently, by considering the economic indices of foreign countries (Iraq, Kuwait, Russian Federation, Saudi Arabia, China, Qatar, Yemen, Belgium, United Arab Emirates, Egypt, Jordan, Libya, Ukraine, Netherlands, Norway, Afghanistan, Germany, Azerbaijan, Iran, United Kingdom, Sweden, Kazakhstan and Other Countries) which demand housing from Turkey, monthly panel data set has been formed for the years 2015–2016. Variables Monthly Number of House Sales to Foreigners by National Nation, European Currency, Gold Bullion, Economic Confidence Index, Consumer Confidence Index, Real Sector Confidence Index, Service Sector Confidence Index, Retail Trade Confidence Index are determined as the shadow variable that takes the value of 1 for investment purpose provinces and 0 for the others, and are described in detail in Tables 18.2 and 18.3.

Table 18.1 Turkey's economic data for expats in demand for housing in Turkey

KT	Housing demand (monthly number of residential sales in overseas settlements 48 months 2013–2016)
EURO	European currency/ Turkish Lira (exchange rate) monthly average 2013–2016
KALTIN	Gold ingot selling gold price (TL/Gr) 2013–2016
D1	Dummy variable that takes value 1 for investment provinces and 0 for others
ITHALAT	General Imports (monthly foreign trade quantity indices by international standard trade classification 2013–2016)
EGE	Economic confidence Index (2013–2016)
TUGE	Consumer confidence Index (2013–2016)
REGE	Real sector confidence Index (2013–2016)
ISGE	Construction sector confidence Index (2013–2016)
ISTHEN	Employment index (seasonally and calendar adjusted index) trade and service indices and change rates (2013–2016)
UBIC	Paid employees (arriving citizens by general working status-residing in Turkey-2013–2016) TSI, citizen entry survey

Table 18.2 Located in housing demand from Turkey, foreign country economic data

KT	Monthly housing sales to foreigners by nationality (24 months 2015–2016)
EURO	European currency/Turkish Lira (exchange rate) average monthly 2015–2016
KALTIN	Gold ingot (selling gold price (TL/Gr) 2015–2016
D1	Dummy variable that takes value 1 for investment provinces and 0 for others
ITHALAT	General imports (monthly foreign trade quantity indices by international standard trade classification 2013–2016)
EGE	Economic confidence index (2015–2016)
TUGE	Consumer confidence index (2015–2016)
REGE	Real sector confidence index (2015–2016)
HSGE	Service sector confidence index (2015–2016)
PTSGE	Retail trade sector confidence index (2015–2016)
D1	Shadow variable that takes value 1 for investment provinces and 0 for others

Table 18.3 The first ten countries that most housing purchases in the years 2015–2016 from Turkey

Country	2015 (Piece)	Country	2016 (Piece)	Rate of change (%)
England	4.552	Iraq	3.726	−15.5
Iraq	4.407	Britain	2.556	−43.8
Russia	2.377	Saudi Arabia	1.827	−20.3
Kuwait	2.299	Kuwait	1.749	−23.9
Saudi Arabia	2.292	Afghanistan	1.623	*
Germany	1.280	Germany	1.474	15.2
Azerbaijan	864	Russia	1.449	−39.0
United Arab Emirates	329	Azerbaijan	724	−16.2
Train	279	Lebanon	191	*
USA	263	TRNC	180	*

Source Eva real estate appraisal (2016)

Clustering analysis was performed to both panel data sets by using k-prototype algorithm. By using cluster validity indexes, the appropriate numbers of clusters were determined and similar units were obtained in housing demand. Cluster validity index values are given in Tables 18.4 and 18.5.

As shown in Table 18.4, for foreign resident data set requesting housing in Turkey, while cluster validity indices of Dunn (0.7467), Ptbiserial (0.7640), Silhouette (0.7733), Thai (0.5965) indices propose the optimal number of cluster as 2, Cindex (0.0111), Gamma (0.9348), Gplus (0.0093) index values recommend as 6. Considering that the number of clusters envisaged for the study will be 6, the number of appropriate clusters is taken as 6. The distribution of provinces in clusters for the number of clusters 6;

1st Cluster: Antalya, Istanbul.

Table 18.4 Cluster validity index values for non-resident data sets in the demand for housing in Turkey

Number of clusters	2	3	4	5	6	7
Cindex	0.0198	0.0224	0.0136	0.0140	**0.0111**	0.0154
Dunn	**0.7467**	0.3820	0.4915	0.3914	0.2132	0.6394
Gamma	0.8674	0.8484	0.8666	0.8831	**0.9348**	0.9289
Gplus	0.0298	0.0349	0.0209	0.0163	**0.0093**	0.0102
McClain	0.2517	0.2352	0.2102	0.2083	0.1808	**0.1380**
Ptbiserial	**0.7640**	0.5144	0.4748	0.4057	0.3336	0.3312
Silhouette	**0.7733**	0.6529	0.6529	0.5859	0.4877	0.5407
Tau	**0.5965**	0.5829	0.5788	0.3775	0.4480	0.3823

Table 18.5 Cluster validity index value for foreign countries in the demand for housing data sets from Turkey

Number of clusters	2	3	4	5	6	7	8	9	10
Cindex	0.0658	0.0182	0.0121	0.0085	0.0070	0.0059	0.0056	**0.0037**	0.0070
Dunn	0.0921	0.1698	0.1259	0.1474	0.1995	0.3361	0.2151	**0.3495**	0.2776
Gamma	0.8012	0.9445	0.9232	**0.94784**	0.9254	**0.94781**	0.9127	0.9333	0.9113
Gplus	0.0501	0.0122	0.0143	0.0130	0.0095	0.0097	0.0044	**0.0041**	0.0050
McClain	0.2668	0.1510	0.1291	0.1155	0.0991	0.0836	0.0840	0.0843	**0.0686**
Ptbiserial	**0.6200**	0.6026	0.5235	0.4782	0.4236	0.3740	0.3522	0.2992	0.3375
Silhouette	0.7064	**0.7707**	0.7471	0.6934	0.6793	0.6043	0.6441	0.6491	0.5725
Tau	0.5714	**0.6301**	0.5656	0.5035	0.4882	0.4508	0.4168	0.3919	0.3936

2nd Cluster: Trabzon.
3rd Cluster: Aydin, Bursa, Other Provinces, Mugla, Yalova.
4th Cluster: Ankara.
5th Cluster: Izmir.
6th Cluster: Sakarya.

as obtained. Comment of Clustering for External Residents;

Antalya is one of Turkey's major cities for tourism. Due to its comfortable living conditions, cultural and historical richness, nature and climate, it attracts the attention of investors from abroad and within the country. Antalya is one of the most immigration receiving provinces and its economy depends on tourism. However, expatriates invested their years of accumulation into residences in their respective cities. Now, implementation of attractive mortgage conditions in Turkey, which are widespread abroad, has caused the Turks to move to Istanbul.

TurkStat's research reveals that Turks living abroad invest in closed sites in Istanbul. Expatriates highlight Antalya and Istanbul in terms of housing investment, which provide both job opportunities and quality living standards, and enable them to be examined in the same cluster. Trabzon, Ankara, Izmir and Sakarya provinces are examined in a cluster by themselves in terms of housing demand. The most important reasons why the expatriate Turks prefer Trabzon in their housing investments are the cool and rainy climate and natural beauties, besides the green nature, housing prices are cheaper compared to provinces such as Istanbul and Antalya, and ease of transportation due to the fact that there is an airport in Trabzon. In addition, the most important reason for the expatriate Turks' housing investments in Ankara is the high potential of Ankara's industry, transportation, industry, education and technical, political, tourism, culture and arts.

On the other hand, Izmir is one of the most preferred cities in terms of housing investment and tourism for expatriate Turks living in Europe. The most important reason why expats prefer 2 + 1 or 3 + 1 house types is that they want to settle in Izmir after their retirement. Thanks to the junctions of the transportation roads belonging to the province of Sakarya and the developing industry, external and internal migrations go on. It is estimated that one million people will live in Sakarya, the shining star of Marmara Region.

Because thermal tourism is important for health, expats demand housing from these places. On the other hand, the real estate sector, which is faced with interest rates, rising costs and exchange rate pressures, Aydın, Bursa, Other Provinces, Mersin, Mugla and Yalova receives high demand from expatriates. Expatriate Turks working abroad and earning foreign exchange make their real estate investments in various cities of Turkey, where they come for holiday.

As shown in Table 18.5, for foreign resident housing demand data set in Turkey, while cluster validity indices of Silhouette (0.7707) and Tau (0.6301) index values propose the optimal number of cluster as three, Cindex (0.0037), Dunn (0.3495) and Gplus (0.0041) indices suggest the optimum number of clusters as 9. Considering that the number of clusters envisaged for the study will be 9, the appropriate number of clusters is taken as 9. The distribution of countries in clusters for 9 clusters;

1st Cluster: Other Countries, Iraq, Kuwait, Russian Federation, Saudi Arabia.
2nd Cluster: China, Qatar, Yemen.
3rd Cluster: Belgium.
4th Cluster: United Arab Emirates, Egypt.
5th Cluster: Jordan.
6th Cluster: Libya, Ukraine.
7th Cluster: Netherlands.
8th Cluster: Norway.
9th Cluster: Afghanistan, Germany, Azerbaijan, Iran, England, Sweden, Kazakhstan as obtained. Comment of Clustering for Foreign Countries;

The housing demand of the five countries in the first cluster and the seven countries in the ninth cluster was determined to be for investment and holiday purposes. In previous years, Afghans who were not seen among countries purchasing houses the most in Turkey, according to the 2016 TSI data, with the purchase of 1623 housings in Turkey are in fifth place in the top 10 most purchases.

Afghans demand housing especially from IstanbulWhen compared with the Germans housing demand of 2015, housing demand interest in Turkey has continued in 2016. In 2016, it ranks second in the list with 1474 units and 649,254 m² purchases. Another country that was.

on the list in the previous year and continued to increase in 2016 is Azerbaijan. Although the number of Azerbaijanis, which ranked ninth in the list in 2015, decreased in number, their purchases in square meters increased. Azerbaijanis ranked fifth with 280,247 m² purchases. It ranks eighth in the list of housing purchases.

According to a research by the Demir, Construction Management Board, border neighborhood of Turkey with Iran it is important from a strategic point of view. Thus, the Iranians are among the most residential area purchasing nations. According to Radikal (2008), the Turkish real estate market since 2003 and it was opened to foreign buyers of the situation, the British property buyers purchase housing from Turkey because of the low prices. Any housing on the Turkish coast can be purchased for as low as 35,000 lb. While the UK is ranked first in the list with 4,552 housing purchases in 2015, it is ranked second with 2,556 housing purchases in 2016.

The similarity of the countries in the second, third, fourth, fifth, sixth, seventh and eighth clusters stems from the fact that housing demand is for investment, accommodation and tourism purposes. In 2015, Iraq was the leading foreign investor with 4 228 units purchase followed by Saudi Arabia with 2704, Kuwait with 2130, Libya with 427, United Arab.

Emirates with 332, Qatar with 277, Egypt with 318, Jordan with 243 and Yemen with 231 houses. Middle Eastern countries mostly buy real estate from Istanbul, Izmit and Yalova provinces.

When we look at the year 2016, Iraqis have 3 thousand 36 houses. The monthly average purchase price of Iraqis, the leader of 2015, was 253. In 2016, Saudi Arabia citizens ranked second with 1886 and Kuwait citizens ranked third with 1744. Accordingly, in 2016, 18 thousand 391 houses were sold to foreigners. The Dutch have purchased 217, the Belgians 198.

18.5   Conclusion

In this study, clustering analysis with k-prototypes algorithm was applied to the economic factors affecting the sale of houses of provincial groups. The main purpose of the study is to classify the provinces and countries showing common feature according to the expectations of the Turks living abroad and foreign countries in the demand for housing in Turkey.

According to the analysis results, the cluster validity indices Cindex (0.0111), Gamma (0.9348) and Gplus (0.0093) as index values found for expatriates who demand housing from Turkey suggest the optimal number of clusters as 6. On the other hand, cluster validity indices Cindex (0.0037), Dunn (0.3495) and Gplus (0.0041) as the index found for foreigners who demand for housing in Turkey suggests the optimal number of clusters as 9. According to the result of clustering; The expectations of Turkish citizens living abroad and foreigners' housing demands have shown similarity.

Expatriates and foreigners mostly want the houses they buy for investment and holiday purposes to be both economical and durable. Expatriates, who prefer summer locations, hometowns and metropolitan cities in the purchase of real estate, revive the housing market. In addition, the high course of foreign exchange, like expatriates, stimulates foreign buyers in housing purchases. Experts in the housing sector state that the VAT advantage granted to foreigners is also granted to Turks residing abroad for more than 6 months with work and residence permits. As a result, the housing market experts predict that there will be significant activity in the Turkish housing market if foreign housing increases and private housing campaigns are provided to expatriate Turks and foreigners.

References

Ahmad A, Iqbal N, Siddiqui R (2018) Determinants of housing demand in urban areas of Pakistan: evidence from the PSLM. Pakistan Dev Rev 57:1–25

Akay Ö, Yüksel G (2018) Clustering the mixed panel dataset using Gower's distance and k-prototypes algorithms. Commun Stat Simul Comput 47:3031–3041

Akay Ö, Yüksel G (2019) Hierarchical clustering of mixed variable panel data based on new distance. Commun Stat Simul Comput 1–16

Bajari P, Kahn ME (2005) Estimating housing demand with an application to explaining racial segregation in cities. J Bus Econ Stat 2:20–33

Bonzo DC, Hermosilla AY (2002) Clustering panel data via perturbed adaptive simulated annealing and genetic algorithms. Adv Comp Syst 5:339–360

Carliner G (1973) Income elasticity of housing demand. Rev Econ Stat 55:528–532

Charrad M, Ghazzali N, Boiteau V, Niknafs A (2012) NbClust package: finding the relevant number of clusters in a dataset. J Stat Softw, Softw

Çelik C, Kıral G (2018a) Kümeleme yöntemiyle konut talebinin incelenmesi: Türkiye il grupları üzerine bir uygulama. Çukurova Üniversitesi Sosyal Bilimler Enstitüsü Dergisi 27:123–138

Çelik C, Kıral G (2018b) Yurtdışı yerleşiklerin ve dış ülkelerin konut taleplerini incelemede panel kümeleme analizi: Türkiye illeri örneği. Çukurova Üniversitesi İktisadi Ve İdari Bilimler Fakültesi Dergisi 22:305–324

Davis M, Oliner S, Peter T, Pinto E (2017) The impact of interest rates on house prices and housing demand: evidence from a quasi-experiment. Working Paper

Do TL, Nguyen TT, Grote U (2019) Livestock production, rural poverty, and perceived shocks: evidence from panel data for Vietnam. J Dev Stud 55:99–119

Goodman AC (1988) An econometric model of housing price, permanent income, tenure choice, and housing demand. J Urban Econ 23:327–353

Hanushek EA, Quigley JM (1980) What is the price elasticity of housing demand? Rev Econ Stat 62:449–454

Huarng KH, Yu THK, Rodriguez-Garcia M (2019) Qualitative analysis of housing demand using Google trends data. Econ Res Ekonomska Istraživanja 409–419

Hou F, Ai K (2015) The empirical research of relationship between consumption and income for Chinese urban residents. Open J Appl Sci 5:251

Huang ZX (1998) Extensions to the k-means algorithm for clustering large datasets with categorical values. Data Min Knowl Disc 2:283–304

Ji J, Pang W, Zhou C, Han X, Wang Z (2012) A fuzzy k-prototype clustering algorithm for mixed numeric and categorical data. Knowl-Based Syst 30:129–135

Ji J, Bai T, Zhou C, Ma C, Wang Z (2013) An improved k-prototypes clustering algorithm for mixed numeric and categorical data. Neurocomputing 120:590–596

Liu X (2019) The income elasticity of housing demand in New South Wales, Australia. Reg Sci Urban Econ 75:70–84

Lu H, Huang S (2011) Clustering panel data. In: SIAM international workshop on data mining held in conjunction with the 2011 SIAM international, conference on data mining. Arizona, USA: IEEE SMCS, pp 1–10

Nguyen CH, Nordman CJ (2018) Household entrepreneurship and social networks: panel data evidence from Vietnam. J Dev Stud 54:594–618

Nunes PJM, Serrasqueiro ZM (2007) Capital structure of Portuguese service industries: a panel data analysis. Serv Ind J 27:549–562

Nunes PJM, Serrasqueiro ZM, Sequeira TN (2009) Profitability in Portuguese service industries: a panel data approach. Serv Ind J 29:693–707

Zheng X, Xia Y, Hui EC, Zheng L (2018) Urban housing demand, permanent income and uncertainty: Microdata analysis of Hong Kong's rental market. Habitat Int 74:9–17

Chapter 19
Evaluation and Optimization of the Treatment Scheme for the Paint Industry Effluents Using Multi-criteria Decision Theory

Seval Sözen, Seyda Duba, Dilber Uzun Ozsahin, Fidan Aslanova, Hüseyin Gökçekuş, and Derin Orhon

Abstract The study was evaluated and optimized the wastewater management for a paint industry. For this purpose, an in-plant survey identified individual waste streams for different processes and characterization and treatability studies were conducted on these segregated waste streams. The resulting treatment scheme involved ten different technical alternatives related to the operational conditions. An evaluation was performed using a multiple-criteria decision tool by analyzing physical footprint and performance of the treatment, chemicals used, the sludge produced, the energy consumed, which defined a sustainable wastewater management involving the optimal treatment configuration.

Keywords Paint industry · Segregated waste streams · Treatability · Wastewater management · Multiple-criteria decision tool

S. Sözen (✉) · S. Duba
Environmental Engineering Department, Faculty of Civil Engineering, Istanbul Technical University, 34469 Maslak, Istanbul, Turkey
e-mail: sozens@itu.edu.tr

D. Uzun Ozsahin
DESAM Institute, Near East University, Nicosia, Turkish Republic of Northern Cyprus, Turkey

Department of Biomedical Engineering, Near East University, Nicosia, Turkish Republic of Northern Cyprus, Turkey

Medical Diagnostic Imaging Department, College of Health Science, University of Sharjah, Sharjah, United Arab Emirates

D. Uzun Ozsahin
e-mail: dilber.uzunozsahin@neu.edu.tr

F. Aslanova · H. Gökçekuş · D. Orhon
Faculty of Civil and Environmental Engineering, Near East University, Nicosia, Turkish Republic of Northern Cyprus, Turkey

D. Uzun Ozsahin et al. (eds.), *Application of Multi-Criteria Decision Analysis in Environmental and Civil Engineering*, Professional Practice in Earth Sciences,
https://doi.org/10.1007/978-3-030-64765-0_19

19.1   Introduction

Paint production is one of the industrial sectors that addresses most to household activities. Paints are continuously utilized worldwide for decoration, protection of building facades, equipment and machinery protection among many other applications. The production scheme is quite diverse and complex depending on the demand, involving pigments, carrier such as resin or other binders, solvents and other additives. These chemicals are partly transferred to the process wastewaters, which pose a serious threat to human life and environment, if not properly handled and treated (Porwal 2015).

Studies on paint industry effluents are rather limited compared to the magnitude of the sector: Among significant studies, Yacout and El-Zahhar (2018) reported on the environmental impact assessment aspect of paint production; Saif et al. (2015) provided useful information on the calculation and estimation of the carbon footprint applicable to paint industry; Dunmade (2012) commented about the recycling aspects and a few studies focused on different treatment approaches (Devletoğlu et al. 2002; Kutluay et al. 2003; Vengris et al. 2012). The earlier work on Lorton (1988) should also be recognized as it reported on waste minimization aspects for paint and allied industries.

While the wastewater is associated with different processes generating individual waste streams in the paint industry, options presented in the literature relied on *end-of-pipe* treatment of combined effluent, perhaps the worst alternative to be considered for paint processing. This study provided a new insight for waste management where segregated waste streams were separately characterized and subjected to treatability analyses.

In this context, the objective of the study was to evaluate and optimize the treatment scheme by considering 10 different technical aspects related to the operational conditions for a paint industry. The evaluation was performed using a *multiple-criteria decision tool* by analyzing physical footprint and performance of the treatment, chemicals used, the sludge produced, the energy consumed.

19.2   Materials and Methods

19.2.1   Process Description

The paint industry was located in Çayırova-Gebze within an Organized Industrial District in Turkey The production activity in the plant was conducted on a 3500 m^2 closed area and mainly consisted of oil-based paint, plastic paint, coating, paint thinner, varnish, and auxiliary paint materials amounting to a yearly production of 25,000 ton.

The capacity was mostly distributed among paint (79%), thinner (13%) and coatings, varnish and other auxiliary paint materials (8%). Alkyd resin was also produced as an intermediate product for the production of ultimate paint and other products at an amount of 6500 ton/year.

The treatment requirement defined for the plant was limited with the "Indirect Discharge Criteria of Industrial Wastewaters" before being discharged to the current common wastewater treatment plant located in the district. The treatment scheme of the plant should provide an effluent quality having no harmful effect on the common wastewater treatment plant.

19.2.2 Wastewater Characteristics

The predominant wastewaters sourced by the plant were categorized in four different types: Water-based paint wastewater and alkyd resin wastewater as process wastewaters; cooling water and domestic wastewater.

The production scheme of water-based paint included first the mixing of the pigments with resins, oils, surfactants and water in a dispersion tank and then the coloring with colorant. The capacity of water-based paint production was 60 ton/day. The amount of wastewater was approximately accounted for 6.5 m^3/day generated intermittently from cleaning of the reactors at a unit production rate of 0.11 m^3/ton water-based paint.

The production of alkyd resin was realized by esterification of polyhydric alcohol, unsaturated aliphatic acids and polybasic acids. The process produced a wastewater of 0.035 m^3 per ton alkyd resin amounting intermittently a total of 1.5 m^3/day.

Cooling system used in the plant was a closed flux generator wasting intermittently an amount of wastewater for 3 m^3 daily. Domestic wastewater sourced by 250 employees was theoretically calculated as 14 m^3/day by considering a personal consumption of 56 L/cap.day. Four wastewater sources resulted in a total amount of 25 m^3/day.

A long-term conventional characterization of 45 days was conducted on process wastewaters and cooling water to outline the difference in characteristics. The experiments were run on composite samples collected from day-time shift. The average values on different parameters were shown in Table 19.1 together with the flowrates. The process wastewaters were tested for only COD, TSS and oil-grease.

The characteristics were then evaluated in a way to treat the effluents with similar characteristics separately. A most striking result in characterization was appeared as the strong characteristic of alkyd resin wastewater in terms of COD. The same property was also reflected by water-based paint wastewaters for TSS and oil-grease parameters.

Table 19.1 Wastewater characterization

Wastewater sources	Flowrate (m³/day)	pH	COD (mg/L)	TSS (mg/L)	Oil-grease (mg/L)	Detergent (mg/L)	NH₄⁺-N (mg/L)
Process wastewaters							
Water-based paint	6.5	7.2	20,000	7000	450	150	10
Alkyd resin	1.5	2	50,000	20	25	NA	NA
Cooling water	3	7	100	150	NA	NA	NA
Domestic wastewater	14	7.1	400	100	150	6	11

NA Not available

19.2.3 Treatability Studies

Treatability studies were primarily designed for testing the most appropriate treatment scheme for main wastewater sources aiming the best solution if applied only to process wastewaters or a combination of different effluents before being discharged to the common wastewater treatment plant in the Organized Industrial District. In this context four treatment schemes were proposed for the different streams, involving a sequence of plain settling, chemical settling and an equalization tank for intermittent process flows prior to biological treatment as defined for the most of the industrial sectors. The alternatives allocated for different wastewater combinations were summarized in Table 19.2. The experiments for different alternatives were run with the flow-weighted composite samples.

The first alternative related to process wastewaters consisted of only a plain settling for 4 h. The unusual long settling time was a trial of understanding if a chemical treatment step could be omitted before biological treatment or not. The plain settled effluent was then mixed with cooling water and domestic wastewater in an equalization tank prior to biological treatment.

In the second alternative, the plain settling time was selected as the half of the first alternative but the treatment was enriched with the chemical treatment step. The

Table 19.2 Treatment alternatives

Alternatives	Plain settling (2 hours)	Plain settling (4 hours)	Chemical settling	Equalization tank	Biological treatment
1. Alternative	–	P	–	P + C + D	P + C + D
2. Alternative	P	–	P	P + C + D	P + C + D
3. Alternative	–	–	P + C	P + C + D	P + C + D
4. Alternative	P + C + D	–	P + C + D	–	P + C + D

P: Process wastewaters, C: Cooling water, D: Domestic wastewater supernatant was then mixed with the rest of the wastewater sources and directed to biological treatment.

In the third alternative the process wastewaters combined with cooling water were first applied to chemical treatment without being plain settled and then mixed with domestic wastewater in the equalization tank before the biological treatment.

In the fourth alternative all wastewaters generated from the plant were first plain settled for 2 h and then treated chemically prior biological treatment.

Plain settling was conducted in Imhoff cones of 1 L volume, settling was observed for 2 h. Settling time was extended to 4 h in the first alternative since no chemical treatment was applied. The performance of the plain settling was determined by measuring COD and TSS.

Chemical treatment tests were implemented using a laboratory jar test equipment. Considering that all process effluents contained metal components, pH value was set to the optimum value of 9 from the original pH with caustic lime. All jar test experiments were conducted applying a sequence of 1 min stirring at 120 rpm for coagulation, 15 min stirring at 30 rpm for flocculation and 30 min settling. $FeSO_4$ and $FeCl_3$ were used as coagulants together with the nonionic and anionic polyelectrolytes to increase the precipitation yield.

$FeCl_3$ and $FeSO_4$ were used as coagulants in chemical treatment as summarized together with their dosages in Table 19.3. pH was adjusted with the addition of caustic lime up to 9–9.5. Only the third alternative was practiced with $FeCl_3$ at 1200 mg/L but with two different polyelectrolyte addition. The chemical treatment in the second and fourth alternatives were conducted with both coagulants at a dosage range of 500–800 mg/L.

Biological treatability studies were conducted in fill and draw activated sludge reactors. The volume of the reactors was selected as 2 L. The reactors were operated at a sludge age of 6 day and a hydraulic retention time of 1 day. Nitrogen and phosphorus requirement for biological activity were supplied by the addition of domestic wastewater.

19.2.4 Analytical Procedures

All conventional parameters except COD were analyzed in accordance with Standard Methods (APHA 2005) COD measurements were conducted by following the procedure in ISO 6060 (1986). pH estimation was conducted as defined in ISO 10523 (2008) by using a digital pH/millivolt meter. Treatability studies and all analysis were performed at room temperature.

Table 19.3 Chemical
treatment procedures for
different alternatives

Alternatives	pH	FeCl$_3$ (mg/L)	FeSO$_4$ (mg/L)
2. Alternative	9.5	500	500
		600	600
		700	700
		800	800
3. Alternative	9	1200[a]	–
		1200[b]	–
4. Alternative	9.5	500	500
		600	600
		700	700
		800	800

[a]With 3 mg/L nonionic polyelectrolyte
[b]With 1 mg/L anionic polyelectrolyte

19.2.5 Multi-criteria Decision Making

Visual PROMETHEE (Preference Ranking Organization Method for Enrichment Evaluation) software was used for Multi-criteria decision making (MCDM). The methodology, PROMETHEE was developed by Brans as one of the well accepted MCDM techniques (Zhaoxu and Min 2010). The advantage of the method was that it did not require any normalization of the scores. Only weighting should defined separately as the weighting techniques was not included in the methodology.

The software necessitated a data matrix including the scores for all criteria defined for treatment alternatives. The scores ranged from 1 to 5, where 1 indicated the worst case, 5 the best. The criteria in data matrix could be weighted considering their importance level. The method used preferential functions to decide the best solution depending on the criteria. Gaussian approach was utilized among 6 different preferential functions. At the end of calculations, the value of highest net outranking flow was accepted as the best decision among the alternatives.

In this study the treatment alternatives were evaluated regarding 10 different decisive parameters (criteria), such as physical footprint of the treatment, use of chemical, amount of sludge originating from the treatment, removal efficiencies on the basis of COD, TSS and metals, and finally energy consumption. The scoring was prepared considering the treatability studies and general informative data. All criteria were weighted as 1 with the understanding that they had the same importance level.

19.3   Results and Discussion

19.3.1   Plain Settling

Plain settling experiments were focused to evaluate the behavior of the settleable organics under two different settling time. The process wastewaters were separately applied to 2 and 4 h settling to understand the effect of settling time on the performance of COD and TSS removal. In 2 h settling the removal efficiency was measured as 35% for COD and 85% for TSS, where they only be increased up to 40% and 90% when the settling time was expanded to 4 h, respectively. Regarding the observation that the doubling in settling time did not reflect any remarkable difference for process wastewaters, the third and fourth alternatives were only applied to 2 h settling coupled with chemical treatment. The performance of 2 h settling for other alternatives was noticed as the same obtained for the process wastewaters.

19.3.2   Chemical Treatment

Chemical treatment was designed first for process wastewaters as in the second alternative and then a combination with cooling water as in the third alternative and finally with the addition of domestic wastewaters as in the fourth alternative.

The performance of chemical treatment if only applied to process wastewaters indicated that 700 mg/L $FeSO_4$ resulted a better quality in terms of TSS (90%) and COD (70%) than other dosages and $FeCl_3$ application. The results indicated that TSS and COD removal together with the metal removal were satisfactory to be discharged into the common wastewater treatment plant.

The combination of process wastewater and cooling water was chemically treated by using only a single coagulant $FeCl_3$ at a higher dosage of 1200 mg/L $FeCl_3$ together with two different polyelectrolytes. The reason using higher dosage was directly due to not applying plain settling. 1 mg/L anionic polyelectrolyte was more effective than 3 mg/L nonionic polyelectrolyte, with a removal yield of 95% for TSS, 70% for COD and 90% for oil-grease.

The results related to chemical precipitation of all wastewaters clearly indicated that best removal was again achieved with $FeSO_4$ application, but at a slightly higher dosage (800 mg/L) than in the second alternative as a consequence of the increased amount of wastewater to be treated. The removal efficiencies were observed to be 95% for TSS, 70% for COD and 90% for oil and grease. Table 19.4 shows the performance obtained from chemical treatment by the application of different coagulants.

A general evaluation of the pretreatment step prior biological treatment revealed that only a COD removal of 38% and a TSS removal of 86% were achieved in the first alternative, whereas these removal efficiencies were almost doubled for COD removal (78%) and slightly higher for TSS removal (95%) in the second alternative.

Table 19.4 Performance of chemical treatment for different alternatives

Alternatives	FeCl$_3$			FeSO$_4$		
	Dosage (mg/L)	COD removal (%)	TSS removal (%)	Dosage (mg/L)	COD removal (%)	TSS removal (%)
2. Alternative	500	60	80	500	65	82
	600	62	82	600	68	85
	700	63	85	700	70	90
	800	65	87	800	71	90
3. Alternative	1200[a]	68	90	–	–	–
	1200[b]	70	95	–	–	–
4. Alternative	500	61	84	500	63	85
	600	63	85	600	65	88
	700	64	88	700	65	90
	800	67	90	800	70	95

[a]With 3 mg/L nonionic polyelectrolyte
[b]With 1 mg/L anionic polyelectrolyte

The third alternative performed a lower removal yield for COD (68%) and TSS (92%) when compared to the second alternative. The most suitable pretreatment scheme seemed to be the fourth alternative where all wastewater sources were applied to plain settling and chemical settling. This alternative ended up with a COD removal efficiency of 78% and a TSS removal of 95%.

Figure 19.1 prepared to outline the performance of each treatment unit in treatment schemes.

19.3.3 Decision Matrix

Starting from the treatability oriented experimental results and general informative data a decision matrix was completed as shown in Table 19.5. All issues related to the size, performance of the treatment units, chemical use, energy consumption and sludge generated in the treatment were comparatively evaluated for the preparation of the matrix and a scoring was proposed by weighting each alternative.

The pretreatment footprint was weighted as 5 (the best score) for the first and third alternatives since both were missing either plain settling or chemical treatment in comparison to second and fourth alternatives. The worst score was devoted to the fourth alternative relying on the fact that the footprint of both plain settling and chemical treatment was higher due to the higher amount of wastewater treated.

Best scouring was attributed to the first alternative with no chemical treatment to reflect the chemical use in chemical treatment. The other allocations were distributed considering the experimental treatability studies.

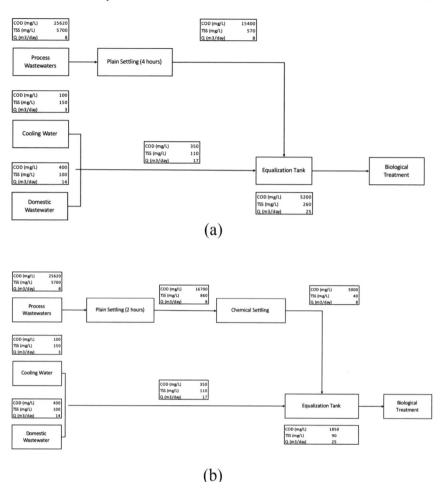

Fig. 19.1 Schematic representation of each alternative, **a** 1. alternative, **b** 2. alternative, **c** 3. alternative, **d** 4. alternative

Comparative evaluation of the sludge generation from plain settling clearly indicated that the third alternative without plain settling had the highest score, and the scoring decreased with the increase of the flowrate to be treated. Similar comparison with the same understanding was completed for the sludge generation from chemical settling. In cases where $FeCl_3$ and $FeSO_4$ were applied at the same dosages the amount of sludge sourced by the use of $FeCl_3$ produced a higher amount of sludge (Balik and Aydin 2016). Special attention should be given to the fourth alternative with the lowest score where domestic wastewater was also chemically treated with other process wastewaters.

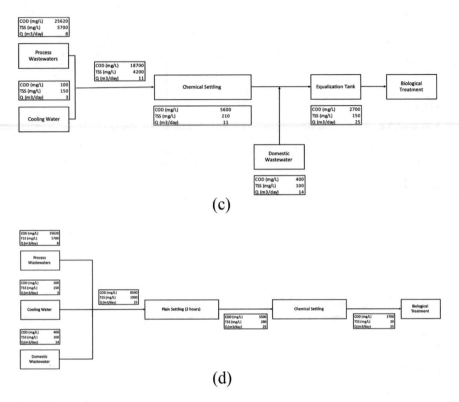

Fig. 19.1 (continued)

Results of the experimental studies were used to define the scores for COD and TSS removal. Metal removal was assumed to be almost removed in chemical treatment. Only the first alternative without chemical treatment was scored with a low number.

The scoring for biological treatment footprint yielded the worst for the first alternative with only plain settling and no chemical treatment application. The score was held as the lowest with the expectation that the biological system could be overloaded in comparison to the other alternatives with chemical treatment. The best scoring was presented to the fourth alternative where the whole amount of wastewater run through plain settling and chemical treatment decreasing the organic load to some extent prior biological treatment. The biological sludge generated in the biological treatment was expected to be highest in the first alternative, that was the reason of being scored as the lowest. The fourth alternative including all wastewaters produce the effective amount of biological sludge since the settleable solids in plain settling and particles suitable for chemical treatment were removed before the biological treatment step. Evaluation on energy consumption depended on the oxygen consumption in biological treatment. The alternatives where the organic loads were decreased by plain settling and/or chemical precipitation received a higher score; on the contrary

Table 19.5 Decision matrix

Alternatives	Pretreatment					Biological treatment						
	Pretreatment Footprint	Chemical use	Sludge from plain setting	Sludge from chemical setting	COD removal	TSS removal	Metal removal	Biological treatment footprint	Energy consumption	Sludge from biological treatment
1. Alternative	5	5	3	5	1	1	2	1	1	1
2. Alternative	4	3	2	3	4	4	5	4	4	4
3. Alternative	5	2	5	1	2	3	5	2	2	2
4. Alternative	3	1	1	3	5	5	5	5	4	4

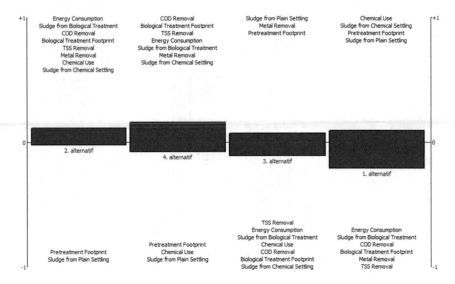

Fig. 19.2 Visual PROMETHEE results

the first alternative with only plain settling loaded higher organics to the biological treatment that should increase the oxygen consumption.

Based on the decision matrix net outranking flow values were calculated using Visual PROMETHEE and the results were illustrated in Fig. 19.2. The visual PROMOTHEE scenario offered a total net flow of outranking ranging from positive to negative. The range from the net flow point to the positive outranking flow presented the progressive performance of alternatives, where the opposite case, to the negative outranking flow represented the descending performance. The ranking of the alternatives calculated were given in Table 19.6.

Table 19.6 identified that the second alternative was in the first place with highest net flow of 0.0885. This alternative seemed more advantageous (approximately %12) than the close follower fourth alternative. The third and the first alternatives were evaluated as the non-feasible treatment alternatives for the paint industry.

Table 19.6 Visual PROMETHEE ranking results

Rank	Phi	Phi+	Phi−
2. Alternative	0.0885	0.1258	0.0372
4. Alternative	0.0786	0.1564	0.0779
3. Alternative	−0.0458	0.0766	0.1224
1. Alternative	−0.1213	0.0892	0.2105

19.4 Conclusion

Multi-criteria decision making (MCDM) analysis provides a guidance to the decision makers to discover the most suitable and optimum solution to the problem in question. The problems related to environmental sector have a wide spectrum of criteria having sometimes conflict with each other.

In this context, the study reflected this new approach for the evaluation and optimization of the wastewater management for a paint industry; It involved an in-plant survey identifying individual waste streams for different processes. Characterization and treatability studies were conducted on these segregated waste streams. The resulting treatment scheme involved ten different technical alternatives related to the operational conditions. An evaluation was performed using a *multiple-criteria decision tool* by analyzing physical footprint and performance of the treatment, chemicals used, the sludge produced, the energy consumed, which defined a sustainable wastewater management involving the optimal treatment configuration.

References

APHA A (2005) Standard methods for the examination of water and wastewater. WEF 21:258–9

Balik ÖY, Aydin S (2016) Coagulation/flocculation optimization and sludge production for pretreatment of paint industry wastewater. Desalin Water Treat 57(27):12692–12699

Devletoğlu O, Philippopoulos C, Grigoropoulou H (2002) Coagulation for treatment of paint industry wastewater. J Environ Sci Heal A 37(7):1361–1377

Dunmade I (2012) Recycle or dispose off? Lifecycle environmental sustainability assessment of paint recycling process. Resour Environ 2(6):291–296. https://doi.org/10.5923/j.re.20120206.07

International Standards Organization (ISO) (1986) Water quality-determination of the chemical oxygen demand-ISO 6060

ISO 10523:2008 (2008) Water quality—determination of pH

Kutluay G, Germirli-Babuna F, Eremektar G, Orhon D (2003) Treatability of water-based paint industry effluents. Fresenius Environ Bull 13(10):1057–1060

Lorton GA (1988) Hazardous waste minimization: Part III Waste minimization in the paint and allied products industry. Japca 38(4):422–427

Porwal T (2015) Paint pollution harmful effects on environment. Int J Environ, Probl, p 3

Saif S, Feroz A, Khan MA, Akhtar S, Mehmood A (2015) Calculation and estimation of the carbon footprint of paint industry. Nat Environ Pollut Technol 14(3):633–638

Vengris T, Binkiene R, Butkiene R, Ragauskas R, Stoncius A, Manusadzianas L (2012) Treatment of water-based wood paint wastewater with Fenton process. Chemija 23(4):263–268

Yacout D, El-Zahhar MA (2018) Environmental impact assessment of paints production in Egypt. In: Proceedings—4th international conference of biotechnology, environment and engineering sciences ICBE 2018, pp 60–65

Zhaoxu S, Min H (2010) Multi-criteria decision making based on PROMETHEE method. In: 2010 international conference on computing, control and industrial engineering 1: 416–418

Printed in the United States
by Baker & Taylor Publisher Services